237

Advances in Polymer Science

Advances in Polymer Science

Recently Published and Forthcoming Volumes

Enzymatic Polymerisation
Volume Editors: Palmans, A.R.A., Heise, A.
Vol. 237, 2010

High Solid Dispersion
Volume Editor: Cloitre, M.
Vol. 236, 2010

Silicon Polymers
Volume Editor: Muzafarov, A.
Vol. 235, 2011

Chemical Design of Responsive Microgels
Volume Editors: Pich, A., Richtering, W.
Vol. 234, 2010

Hybrid Latex Particles – Preparation with Emulsion
Volume Editors: van Herk, A.M., Landfester, K.
Vol. 233, 2010

Biopolymers
Volume Editors: Abe, A., Dušek, K., Kobayashi, S.
Vol. 232, 2010

Polymer Materials
Volume Editors: Lee, K.-S., Kobayashi, S.
Vol. 231, 2010

Polymer Characterization
Volume Editors: Dušek, K., Joanny, J.-F.
Vol. 230, 2010

Modern Techniques for Nano- and Microreactors/-reactions
Volume Editor: Caruso, F.
Vol. 229, 2010

Complex Macromolecular Systems II
Volume Editors: Müller, A.H.E., Schmidt, H.-W.
Vol. 228, 2010

Complex Macromolecular Systems I
Volume Editors: Müller, A.H.E., Schmidt, H.-W.
Vol. 227, 2010

Shape-Memory Polymers
Volume Editor: Lendlein, A.
Vol. 226, 2010

Polymer Libraries
Volume Editors: Meier, M.A.R., Webster, D.C.
Vol. 225, 2010

Polymer Membranes/Biomembranes
Volume Editors: Meier, W.P., Knoll, W.
Vol. 224, 2010

Organic Electronics
Volume Editors: Meller, G., Grasser, T.
Vol. 223, 2010

Inclusion Polymers
Volume Editor: Wenz, G.
Vol. 222, 2009

Advanced Computer Simulation Approaches for Soft Matter Sciences III
Volume Editors: Holm, C., Kremer, K.
Vol. 221, 2009

Self-Assembled Nanomaterials II
Nanotubes
Volume Editor: Shimizu, T.
Vol. 220, 2008

Self-Assembled Nanomaterials I
Nanofibers
Volume Editor: Shimizu, T.
Vol. 219, 2008

Interfacial Processes and Molecular Aggregation of Surfactants
Volume Editor: Narayanan, R.
Vol. 218, 2008

New Frontiers in Polymer Synthesis
Volume Editor: Kobayashi, S.
Vol. 217, 2008

Polymers for Fuel Cells II
Volume Editor: Scherer, G.G.
Vol. 216, 2008

Polymers for Fuel Cells I
Volume Editor: Scherer, G.G.
Vol. 215, 2008

Photoresponsive Polymers II
Volume Editors: Marder, S.R., Lee, K.-S.
Vol. 214, 2008

Enzymatic Polymerisation

Volume Editors: Anja R.A. Palmans
Andreas Heise

With contributions by

G.M. Guebitz · A. Heise · K. Loos · D. Mecerreyes
E. Ochoteco · A.R.A. Palmans · S. Roy
R.V. Ulijn · M.A.J. Veld · J. van der Vlist

Springer

Editors
Dr. Anja R.A. Palmans
Department of Chemical Engineering
& Chemistry
Eindhoven University of Technology
PO Box 513
5600 MB Eindhoven
The Netherlands
a.palmans@tue.nl

Dr. Andreas Heise
School of Chemical Sciences
Dublin City University
Glasnevin
Dublin 9
Ireland
andreas.heise@dcu.ie

ISSN 0065-3195 e-ISSN 1436-5030
ISBN 978-3-642-16375-3 e-ISBN 978-3-642-16376-0
DOI 10.1007/978-3-642-16376-0
Springer Heidelberg Dordrecht London New York

Library of Congress Control Number: 2010936634

Cover design: WMXDesign GmbH, Heidelberg

Printed on acid-free paper

Springer is part of Springer Science+Business Media (www.springer.com)

Volume Editors

Dr. Anja R.A. Palmans
Department of Chemical Engineering & Chemistry
Eindhoven University of Technology
PO Box 513
5600 MB Eindhoven
The Netherlands
a.palmans@tue.nl

Dr. Andreas Heise
School of Chemical Sciences
Dublin City University
Glasnevin
Dublin 9
Ireland
andreas.heise@dcu.ie

Editorial Board

Advances in Polymer Sciences
Also Available Electronically

Advances in Polymer Sciences is included in Springer's eBook package *Chemistry and Materials Science*. If a library does not opt for the whole package, the book series may be bought on a subscription basis. Also, all back volumes are available electronically.

For all customers who have a standing order to the print version of *Advances in Polymer Sciences*, we offer free access to the electronic volumes of the Series published in the current year via SpringerLink.

If you do not have access, you can still view the table of contents of each volume and the abstract of each article by going to the SpringerLink homepage, clicking on "Browse by Online Libraries", then "Chemical Sciences", and finally choose *Advances in Polymer Science*.

You will find information about the

- Editorial Board
- Aims and Scope
- Instructions for Authors
- Sample Contribution

at springer.com using the search function by typing in *Advances in Polymer Sciences*.

Color figures are published in full color in the electronic version on SpringerLink.

Aims and Scope

The series *Advances in Polymer Science* presents critical reviews of the present and future trends in polymer and biopolymer science including chemistry, physical chemistry, physics and material science. It is addressed to all scientists at universities and in industry who wish to keep abreast of advances in the topics covered.

Review articles for the topical volumes are invited by the volume editors. As a rule, single contributions are also specially commissioned. The editors and publishers will, however, always be pleased to receive suggestions and supplementary information. Papers are accepted for *Advances in Polymer Science* in English.

In references *Advances in Polymer Sciences* is abbreviated as *Adv Polym Sci* and is cited as a journal.

Special volumes are edited by well known guest editors who invite reputed authors for the review articles in their volumes.

Impact Factor in 2009: 4.600; Section "Polymer Science": Rank 4 of 73

Preface

The application of nature's toolset is a fast-growing area in several industries, such as food, fine chemicals and polymers. Specifically in vitro enzymatic catalysis has seen a steady increase in (industrial) applications where enzyme catalysis has established itself as an indispensable tool in the synthesis of small molecules. Examples can be found in the production of pharmaceutical intermediates, where biotechnology generates significant turnover and reduces the environmental impact. In recent years, enzyme catalysis was also successfully applied in polymer synthesis. The motivation for using enzymes in polymer synthesis was initially mainly scientific curiosity, but as this technology started to produce results comparable to conventional polymerizations, the potential was recognized. The possibility of making polymers that are not available from conventional methods and their natural character make enzymes a particularly promising catalytic system. The goal of this book is to provide an overview of enzyme catalysis in polymer synthesis structured according to the different enzyme classes used in these reactions. Notably, three of the six enzyme groups have been reported in enzymatic polymerization in vitro, i.e. oxidoreductases, transferases and hydrolases.

Chapter 1 reviews recent advances in the field of biocatalytic synthesis of water-soluble conducting polymers using oxidoreductases. This class of enzymes catalyses the reduction or oxidation of functional groups. Horseradish peroxidase (HRP), obtained from natural and renewable sources, efficiently catalyzes the oxidative polymerization of aniline and phenol-based monomers under benign conditions. In addition, the technologically relevant conjugated polymer poly-3,4-ethylenedioxythiophene (PEDOT) can be accessed using HRP catalysis. This biocatalytic method results in PEDOT materials that show a high electrical conductivity and possess excellent film formation, making it a valuable tool in the preparation of functional polymers.

Chapter 2 summarizes the application of transferases in polymer chemistry. Transferases are enzymes transferring a group from one compound (donor) to another compound (acceptor). Of the three classes of enzymes used in polymer science, transferases are the least frequently applied, which is due to their sensitivity. Nonetheless, several transferases such as phosphorylases and synthases have been found to be effective for catalyzing the in vitro synthesis of polysaccharides and polyesters and well-defined polymers with a variety of architectures have become available.

Enzymes that belong to the class of hydrolases are by far the most frequently-applied enzymes in polymer chemistry and are discussed in Chaps. 3–6. Although hydrolases typically catalyse hydrolysis reactions, in synthetic conditions they have also been used as catalysts for the reverse reaction, i.e. the bond-forming reaction. In particular, lipases emerged as stable and versatile catalysts in water-poor media and have been applied to prepare polyesters, polyamides and polycarbonates, all polymers with great potential in a variety of biomedical applications.

Chapter 3 focuses on the increased understanding in enzymatic strategies for the production of well-defined polymers. A wide variety of (co)polymers has been synthesised and explored in a variety of applications using lipase catalysts. On the other hand, detailed studies also revealed the limitations of the use of lipases: as a result of the monomer-activation mechanism, polymers of low polydispersity and quantitative degree of end-group functionality are difficult to attain.

Chapter 4 shows that the range of polymeric structures from enzymatic polymerization can be further increased by combination with chemical methods. The developments in chemoenzymatic strategies towards polymeric materials in the synthesis of polymer architectures such as block and graft copolymers and polymer networks are highlighted. Moreover, the combination of chemical and enzymatic catalysis for the synthesis of unique chiral polymers is discussed.

Chapter 5 shows that the application of hydrolytic enzymes is a powerful yet mild strategy to directly improve polymer surface properties (i.e. hydrophilicity) or activate materials for further processing. The surface hydrolysis of polyamides (PA), polyethyleneterphthalates (PET) and polyacrylonitriles (PAN) is discussed, as well as the mechanistic details on the enzymatic surface hydrolysis. The mechanistic data, combined with advances in structural and molecular biology, help to explain different activities of closely related enzymes on polymer surfaces.

Finally, Chap. 6 deals with the exploitation of biocatalysis in generating supramolecular polymers, a class of polymers where the monomers are connected via non-covalent bonds. This approach provides highly dynamic and reversible supramolecular structures, inspired by biological polymeric systems found in the intra- and extracellular space. A number of potential applications of enzymatic supramoleular polymerizations are discussed in the context of biomedicine and nanotechnology.

From the present book and the work of numerous researchers in the field, it becomes clear that enzymatic polymerization presents a serious alternative to chemical synthesis. Significant progress has been achieved in recent years in this relatively young polymerization technique. As with all polymerization techniques, enzymatic polymerizations have their advantages and disadvantages. It is clear that many challenges still have to be overcome, even for polymer systems in which the enzymatic processes have significantly advanced, as is the case for polyester synthesis. Nevertheless, enzymes hold significant promises with respect to green polymer chemistry and as an additional tool for the synthesis of functional polymers. However, the advantages of enzymes are not directly transferable from small molecule synthesis to polymers. While much fundamental research has been done in the past and will be needed in the future on this topic, we are approaching a transition towards the development of integrated green processes in polymer science, and enzymatic polymerization is one element of this transition.

Summer 2010 *Andreas Heise and Anja Palmans*

Contents

Adv Polym Sci (2010) 237: 1–19
DOI:10.1007/12_2010_72

Published online: 17 June 2010

Oxireductases in the Enzymatic Synthesis of Water-Soluble Conducting Polymers

Estibalitz Ochoteco and David Mecerreyes

Abstract This chapter reviews recent advances in the field of biocatalytic synthesis of water-soluble conducting polymers. Biocatalysis is proposed as a versatile tool for synthesis of conducting polymers. First, the enzymatic synthesis of conducting polymers and its mechanism is discussed as well as the use of different type of enzymes. Next, we describe the use of a new bifunctional template (sodium dodecyl diphenyloxide disulfonate) in the synthesis of polyaniline as a strategy to improve the water solubility and electrical conductivity in the obtained polymer. The recent development of enzyme-catalyzed polymerization of 3,4-ethylenedioxythiophene (EDOT) in the presence of polystyrenesulfonate is discussed. This method results in PEDOT materials that show an electrical conductivity of $2 \times 10^{-3}\,\mathrm{S\,cm^{-1}}$ and posses excellent film formation ability, as confirmed by atomic force microscopy images. Finally, a simple method for immobilizing horseradish peroxidases in the biocatalytic synthesis of water-soluble conducting polymers is presented. This method is based on a biphasic catalytic system in which the enzyme is encapsulated inside the ionic liquid 1-butyl-3-methylimidazolium hexafluorophosphate, while other components remain in the aqueous phase. The enzyme is easily recovered after reaction and can be reused several times.

Keywords Conducting polymers · Enzymatic polymerization · Horseradish peroxidase · Oxidoreductases · PEDOT · Polyaniline · Synthesis strategies

Contents

E. Ochoteco (✉) and D. Mecerreyes
New Materials Department, CIDETEC-Centre for Electrochemical Technologies, Parque Tecnológico de San Sebastián, Paseo Miramón 196, 20009 Donostia-San Sebastian, Spain
e-mail: eochoteco@cidetec.es; dmecerreyes@cidetec.es

1 Introduction

The field of intrinsically conducting polymers (ICPs) has attracted a great deal of attention due to their interesting electrical and optical properties. Since their discovery in 1977, these materials have been investigated for countless technological applications, such as organic lightweight batteries, microelectronics, optical displays, antistatic coatings, and electromagnetic shielding [1, 2]. These extremely promising materials have traditionally been synthesized by monomer oxidation in the presence of a strong oxidant and show high insolubility and intractability when synthesized [3]. In this context, the use of enzymes as biological catalysts in the synthesis of functional polymers has attracted great interest in recent years as a "green" alternative [4–6]. The approach does not require strong acids or purification steps in the synthesis process. Enzymes can offer environmentally benign reaction conditions [7], a high yield of polymerization, and a superior level of control in regioregularity and stereochemistry, consequently resulting in soluble and processable conducting polymers [8].

This chapter provides a review of some of the most recent research in the field of biocatalytic synthesis. The reviewed papers propose the development of new strategies in the enzymatic biocatalytic approach to polymer synthesis, providing an answer to several problematic issues. Research topics include: (1) the use of new bifunctional templates in enzymatic polymerization, giving as a consequence highly soluble and conducting polymers; (2) achievement of the first enzymatic polymerization of 3,4-ethylenedioxythiophene (EDOT), demonstrating that this green synthesis process can be applied to other technologically interesting polymers apart from polyaniline (PANI); and (3) the employment of an ionic liquid (IL)-based biphasic catalytic system as a mean to recover and reuse the enzyme several times in aniline or EDOT polymerization.

2 Peroxidases in Polymer Synthesis

Peroxidases function as oxidoreductases, which catalyze the oxidation of a donor using hydrogen peroxide as the oxidizing agent. During this process, two water molecules are liberated (Scheme 1). In biocatalytic reactions, a peroxidase enzyme such as horseradish peroxidase (HRP) or palm tree peroxidase, or even oxidases such as glucose oxidase are used as catalyst for the monomer (e.g., aniline) polymerization in an aqueous buffer [10]. During polymerization, the polymer precipitates

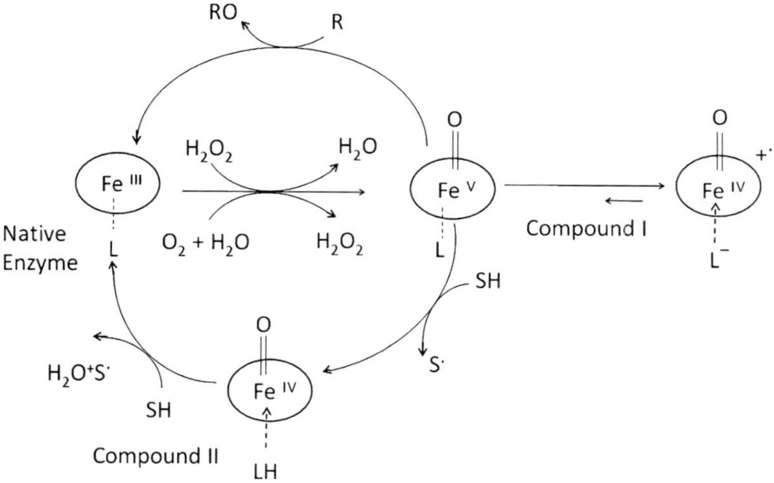

Scheme 1 Catalytic cycle of peroxidase. *L* ligand, *R* reduced compound. (Reprinted with permission from Gross et al. [9]. © 2001, American Chemical Society)

out of the reaction solution, which limits the applicability of this method [11]. Although several strategies have been used to avoid this precipitation, such as modification of monomer structures [12, 13], synthesis in micellar media [14], reverse micelles [15, 16], and interfacial polymerizations [17], the final conductivity of the obtained PANI is poor in most cases. This poor solubility is said to be due to the presence of different structures in the polymer, giving a highly branched instead of a predominantly head-to-tail linear structure, as is formed in the traditional chemical polymerization of aniline. Liu and coworkers [18, 19] successfully improved the conductivity of enzymatically synthesized PANI by using a polymer electrolyte such as poly(sodium *p*-styrenesulfonate) (PSSNa) in the reaction media. This water-soluble polyelectrolyte serves as a linear template. Since PSSNa and aniline have a pKa of 0.70 and 4.63, respectively, at pH close to 4, PSSNa is anionically charged while aniline is cationically charged. Therefore, aniline and PSSNa form an electrostatic complex, which leads to a pre-orientation of the aniline monomers and favors a *para*-directed reaction. This results in a more linear, highly conjugated structure. Consequently, the PANI conductivity is improved by this method. The same effect is obtained with dodecylbenzene sulfonic acid (DBSA), hexadecyltrimethylammonium bromide (HDTMAB) and polyoxyethylene isooctylphenyl ether (PEOPE) as templates. These molecules aggregate to form micelles above a critical concentration (critical micellar concentration, CMC), creating another template configuration for a guided polymerization. However, the use of small molecules such as sodium benzene sulfonic acid (SBS), without the ability to form these templates, does not result in a highly conducting PANI by enzymatic synthesis. Therefore, it seems clear that the pre-orientation of anionic species (via aggregation of small molecules or polymeric electrolytes) is essential to create an adequate environment by (1) providing the necessary counteranions for the doping of the conducting polymer,

(2) improving water solubility, and (3) templating the alignment for monomer molecules during polymerization. The use of surfactants as micelles or as soft templates to homogeneously dispersed the aniline in the aqueous solution continues to be one of the best strategies for obtaining the desired conductive emeraldine form of PANI by peroxidase-triggered polymerization [20, 21]. However, in spite of these recent advances, the electrical conductivity in the final complex remains low for practical applications, and new strategies are still being searched for to attain high conductivity and excellent water solubility.

One of the most important drawbacks of these strategies is the cost of the enzyme. For practical applications, the expensive enzymes must be recovered and reused after the reaction. This is the reasoning behind the well-established strategy of enzyme immobilization on solid supports, which has been applied to HRP [22–26]. HRP has been immobilized on chitosan powder [27], aluminum oxide [28], functionalized polyethylene [29], collagen [30], titanium oxide [31], carbon nanotubes [32], and other templates [22–26]. In fact, new alternative strategies are being studied in many research groups for recycling and reusing the enzyme in the biocatalytic synthesis of polymers.

Finally, it is worth mentioning that biocatalytic synthesis has been successfully applied to PANI as conducting polymer, but it has been hardly extended to other technologically interesting polymers such as polythiophenes [33–36]. Of particular interest, poly(3,4-ethylenedioxythiophene), PEDOT is becoming very successful for commercial applications. The lack of success in the enzymatic polymerization of this material is clearly stated in a recent paper [37], where it is said to be probably due to the high oxidation potential of monomers such as EDOT and pyrrole. Referring to the oxidation potential of the HRP/H_2O_2 pair, it was concluded that EDOT and pyrrole are inappropriate monomers for this enzymatic synthesis.

3 New Bifunctional Templates as a Means to Improve Electrical Conductivity

As previously stated, the use of templates such as DBSA, HDTMAB, and PEOPE allows the formation of well-defined micellar structures in aqueous solution when the template concentration is above its CMC. In a recent publication [38], the polymerization with a bifunctional sodium dodecyl diphenyloxide disulfonate (DODD) as template was proposed to proceed by a micellar mechanism in the same way (Scheme 2). In an aqueous acid solution of DODD and aniline, anilium ions locate at the micellar interface, with benzene parts penetrating into the hydrophobic core of the DODD micelle to form the complex (as illustrated in Scheme 3b). At the concentration of 0.055 mol L^{-1}, a slight turbidity was observed in solution, indicating micellar aggregation. Once the micellar structure is formed and the enzyme is added to the aqueous medium, addition of H_2O_2 triggers the polymerization of anilinium ions around micelles (Scheme 3).

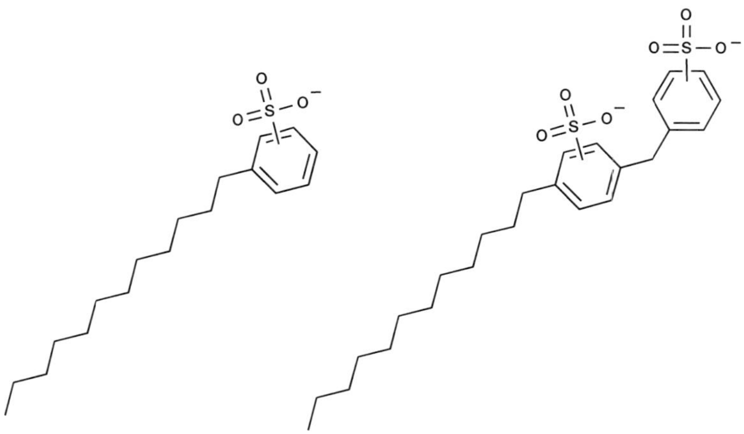

Scheme 2 Chemical structure of DBSA (*left*) and DODD (*right*) molecules. (Reprinted with permission from Rumbau et al. [38]. © 2007, Elsevier)

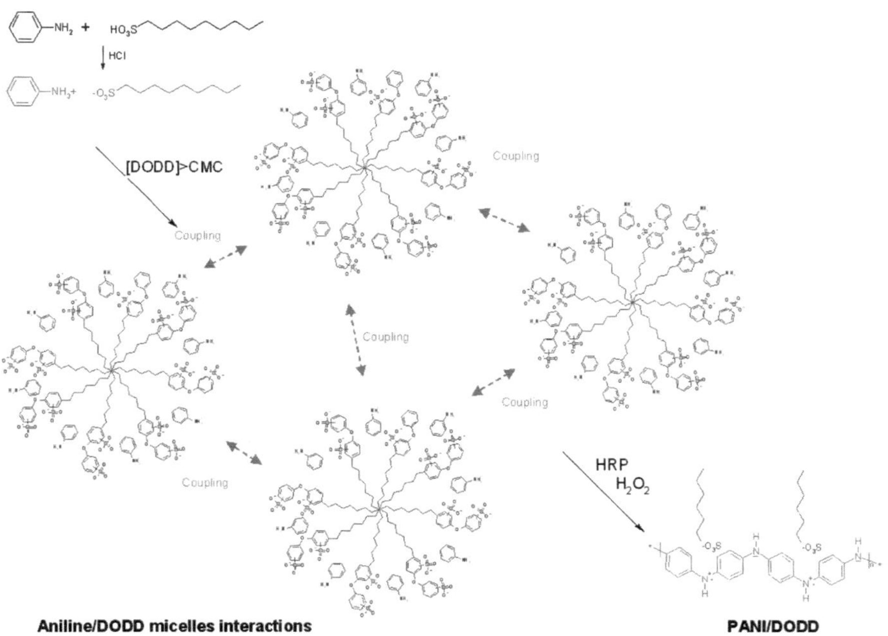

Scheme 3 Polymerization mechanism: aniline–DODD interaction and polymerization. (Reprinted with permission from Rumbau et al. [38]. © 2007, Elsevier)

The bifunctional DODD molecule, due to the presence of two sulfonic groups in its structure, shows higher water solubility than monofunctional surfactants such as DBSA. At acidic pH, sulfonate groups become negatively charged, thereby increasing the total charge per molecule and, consequently, the solubility. This idea

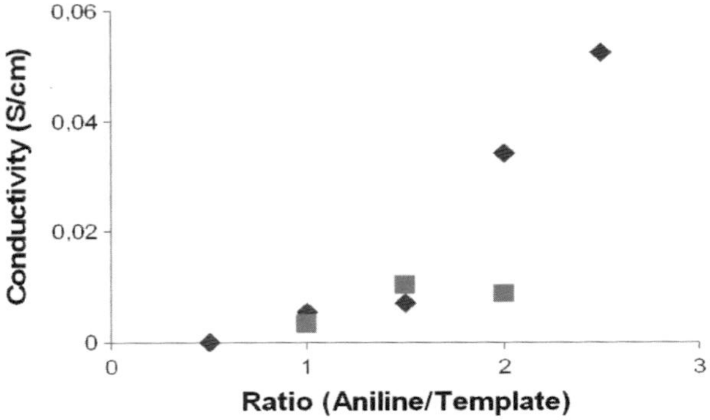

Fig. 1 Influence of aniline to template ratio on conductivity for enzymatic PANI/DODD (*diamonds*) and enzymatic PANI/DBSA (*squares*). (Reprinted with permission from Rumbau et al. [38]. © 2007, Elsevier)

is reflected in the fact that the CMC value for DODD is higher than that for DBSA (7×10^{-3} and 1.6×10^{-3}M, respectively). Furthermore, in a bifunctional surfactant, each micelle is expected to be composed of a lower number of individual molecules than micelles created by monofunctional surfactants and, consequently, to be smaller in size. Light scattering measurement verified this assumption, where micelles with diameters of <4 nm and 5 nm were observed for DODD and DBSA, respectively. It was therefore assumed that, for the same surfactant concentration, there are more micelles formed from a bifunctional surfactant than from monofunctional surfactant at the same concentration.

The content of anionic charges and the anionic template surface for complex formation in the media should thus be higher. The solubility of the polymer complex is explained by the anionic charges on the template, and that a percentage of these anionic charges must remain free to keep the complex in solution. A system with a larger anionic surface would result in higher complexation with anilinium cations on the surface and, consequently, a higher polymerization conversion before precipitation occurred. This higher local anilinium cation concentration should lead to higher molecular weights and, consequently, higher conductivity in the product.

Experimental results (Fig. 1) show that for PANI/DODD and PANI/DBSA synthesized under the same conditions, the conductivity of the PANI/DODD complex is indeed higher than that of the PANI/DBSA. Furthermore, the DODD template allows the use of a higher aniline to template ratio before precipitation occurs. PANI/DBSA complexes show precipitation at ratios close to 2.5:1 whereas PANI/DODD complexes remain in solution even at a ratio of 4:1.

4 First Enzymatic Polymerizations of EDOT

Oxidoreductases such as HRP, obtained from natural and renewable sources, have been known to catalyze the polymerization of aniline and phenol-based monomers under benign conditions and in mixed solvent systems [39–41]. The mechanism for HRP-catalyzed polymerization requires the interaction of the heme-iron active group of the enzyme with the hydrogen peroxide, generating an oxidized heme-iron complex [42]. Then, the oxidized heme-iron complex reacts with the monomer, in a one-electron transfer reaction to give a monomer radical and a modified iron-heme complex. The polymer is formed after consecutive reaction of these monomer radicals. This biocatalytic synthesis process has been successfully applied to PANI as conducting polymer, but rarely to other technologically interesting polymers such as polythiophenes [33–36]. In a work by Bruno et al. [37], this is said to be due to the higher oxidation potential of monomers such as EDOT and pyrrole compared to the catalyst HRP, thus proving them to be inappropriate substrates for this enzymatic approach. Encouraged by the effectiveness of Fe^{2+} as a "green" biomimetic catalyst, the synthesis of water-soluble PEDOT and polypyrrole was performed by the same group using PEGylated hematin (PEG-hematin) as catalyst in presence of PSSNa as template. The EDOT polymerization reaction with PEG-hematin was spectroscopically monitored in an aqueous solution at pH 1.0. The spectral changes observed in Fig. 2 indicate that the monomer did not show significant absorption above 300 nm. However, once H_2O_2 was added, the polymerization process led to the appearance of absorption from 600 to 1200 nm, accompanied by the development of a dark blue color, with a simultaneous increase in the absorption intensity of the peaks over time. The broad band at 700 nm, with a large absorption tail at around 1200 nm, was attributed to the ð–ð* transition in the polymer chain.

The FT-infrared (FTIR) spectrum (Fig. 3) of the polymer showed absorptions at 1342, 1218, and 976 cm^{-1}, all assigned to doping interactions of SPS. The absence of bands at 1600–1800 cm^{-1} was significant and indicated that no overoxidation or ring-opening had occurred. This PEDOT was stable over a fairly high range of temperature (Fig. 3), observing a significant degradation at 110 and 250°C. The initial decrease in weight percent is due to evaporation of bound water (Fig. 4). The electrical conductivity data for the PEDOT (1.10^{-3} S cm^{-1}) are similar to the measured conductivity of commercial PEDOT (Baytron).

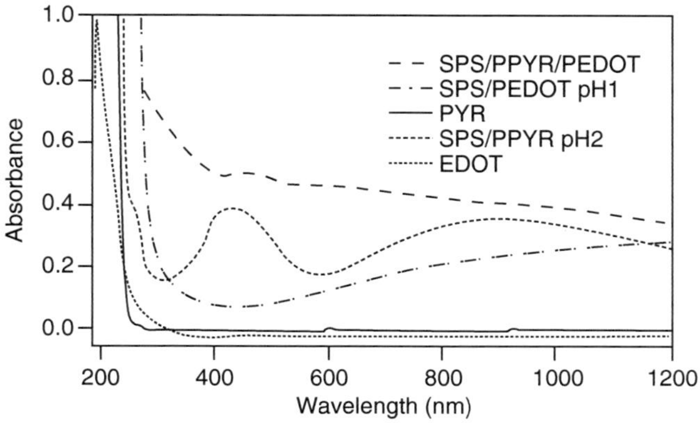

Fig. 2 UV-Vis spectra of monomers EDOT and pyrrole, and polymers PEDOT, polypyrrole and PEDOT-*co*-polypyrrole. (Reprinted with permission from Bruno et al. [37]. © 2006, American Chemical Society)

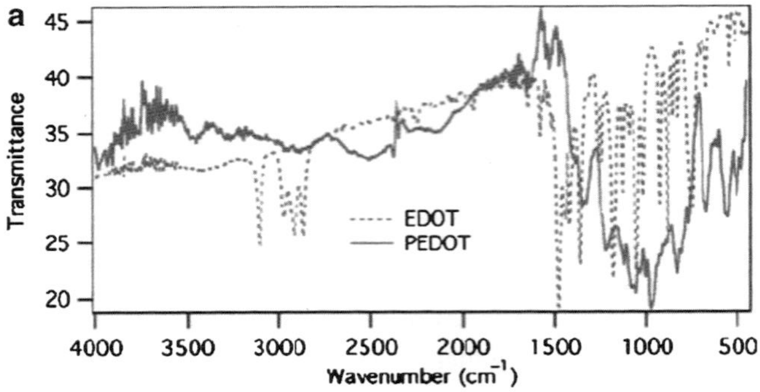

Fig. 3 FTIR spectra of monomer EDOT and polymer PEDOT in the presence of PSSNa. (Reprinted with permission from Bruno et al. [37]. © 2006, American Chemical Society)

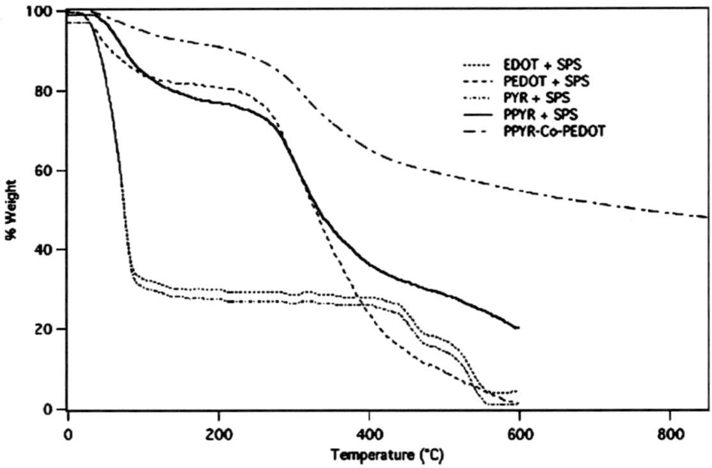

Fig. 4 TGA of the different monomers and polymers. (Reprinted with permission from Bruno et al. [37]. © 2006, American Chemical Society)

In another research laboratory, surprised by the lack of success of other research groups and the previous statements about the impossibility of applying biocatalytic chemistry to polithiophenes and polypyrroles, special attention was paid to the enzymatic polymerization of the EDOT monomer [43]. In this case, the first trials succeeded and a blue-colored polymer solution was obtained after 16 h of reaction (Scheme 4). As is well-known, an acidic reaction medium is suitable to increase the rate of polymerization. Protonic acids and a variety of Lewis acids catalyze the equilibrium reaction of EDOT to the corresponding dimeric and trimeric compounds without further oxidation or reaction [44]. In this work, three different pHs were evaluated (pH = 2, 4, and 6) in order to establish the optimum for adequate synthesis of EDOT. The UV-visible (UV-Vis) spectra for these three reactions are

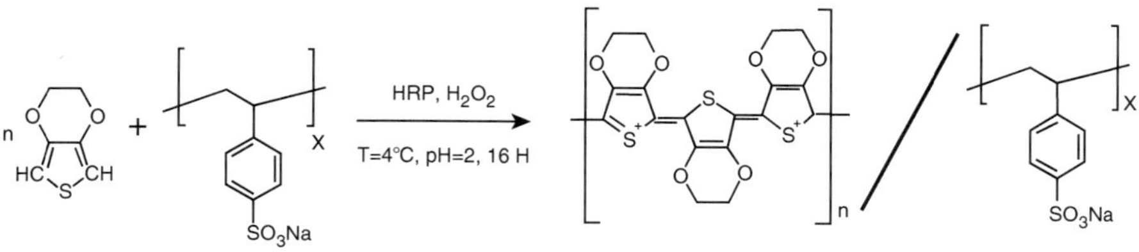

Scheme 4 Mechanism of enzymatic polymerization of EDOT. (Reprinted with permission from Rumbau et al. [43]. © 2007, American Chemical Society)

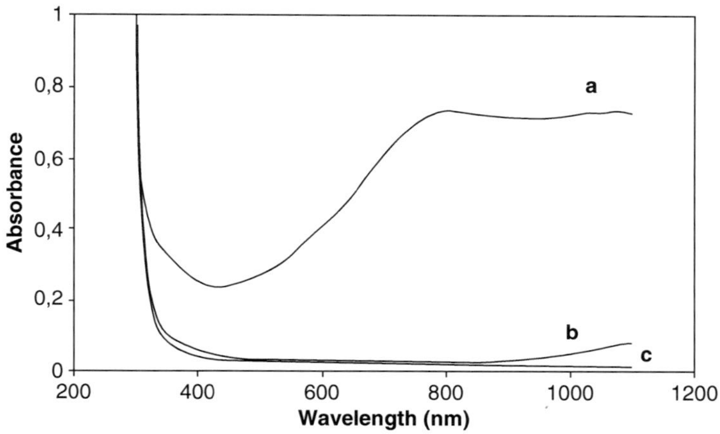

Fig. 5 UV-Vis spectra for PEDOT complexes synthesized for 16 h at (*a*) pH 2, (*b*) pH 4, and (*c*) pH 6. (Reprinted with permission from Rumbau et al. [43]. © 2007, American Chemical Society)

shown in Fig. 5. It can be clearly observed that the only spectra showing the presence of a bipolaron absorption band at 800 nm, associated with PEDOT polymer, was the one obtained at pH 2.

The results demonstrate that the high oxidation potential of the monomer may not be the reason for the lack of polymerization in previous failed attempts. The choice of appropriate working conditions (pH 2, and a polymerization time as long as 16 h) allows the biocatalytic synthesis of PEDOT polymer. On the other hand, it is well known that the activity of HRP decreases abruptly in acidic media (at pH 4, the HRP activity is around 0 after 1 h [2]) but, surprisingly, even though the enzymatic activity should normally be significantly reduced sometime after the first 60 min, the reaction was observed to proceed uninterruptedly throughout the 16 h. The authors thought that, under these strong acidic conditions, the enzyme could be deactivated and the polymerization process could be triggered only by the presence of H_2O_2. In a control experiment performed employing the same conditions except for the absence of HRP, no polymerization was observed. In conclusion, the HRP was assumed to be the catalyst triggering the polymerization reaction and the enhanced enzyme activity due to an excess of EDOT monomer in the media. As the solubility of the monomer is very poor in water, the reaction takes place in a biphasic media, where the HRP is localized more preferentially in the monomer phase. To prove this statement, a mixture of EDOT, template, and enzyme was prepared in acidic water (pH 2). This solution was kept under magnetic stirring for 24 h. Then, a solution of H_2O_2 was added. The reaction proceeded as usual and a conducting water-soluble PEDOT was obtained after an additional 16 h. Thus, as a consequence, the EDOT monomer droplets act as (1) enzyme protectors against deactivation, and (2) monomer feed to keep a constant EDOT concentration in the reaction media. The HRP-catalyzed enzymatic polymerization of EDOT was thus successfully performed at the specific synthesis conditions (low pH, protected enzyme). Also, FTIR spectroscopy confirmed the classical bands of PEDOT at 1698, 1381, 1066, and 940–875 cm^{-1}, and of the PSSNa stabilizers at 1009 and 1128 cm^{-1} (Fig. 6). Furthermore, the electrical conductivity of the films was verified by the four-points probe. Typical values for PEDOT of 2×10^{-3} S cm^{-1} were measured, further confirming the success of our synthetic route.

The obtained PEDOT polymer showed a very good capacity for film formation at a macroscopic level, and very low surface roughness also at a microscopic level, as can be observed in the surface picture of the film obtained by atomic force microscopy (Fig. 7).

Despite these successful polymerizations, these reactions proceed well only at low pH and temperatures, which is restrictive for some applications. Therefore, the most recent work on the biocatalytic polymerization of EDOT reports the use of terthiophene as a radical mediator to synthesize PEDOT/PSSNa under milder pH and at ambient conditions using the enzyme soybean peroxidase (SBP) and PSSNa [44]. SBP belongs to the family of plant peroxidases that can oxidize a wide variety of organic and inorganic substrates using hydrogen peroxide. Although SBP has been reported to have a higher redox potential than HRP, SBP cannot catalyze the polymerization of EDOT under non-acidic conditions. Therefore, the polymerization of EDOT was accomplished in this work by introducing a small amount (<1%) of an oligomeric thiophene such as 2,5-di(2-thienyl)thiophene [terthiophene] in the

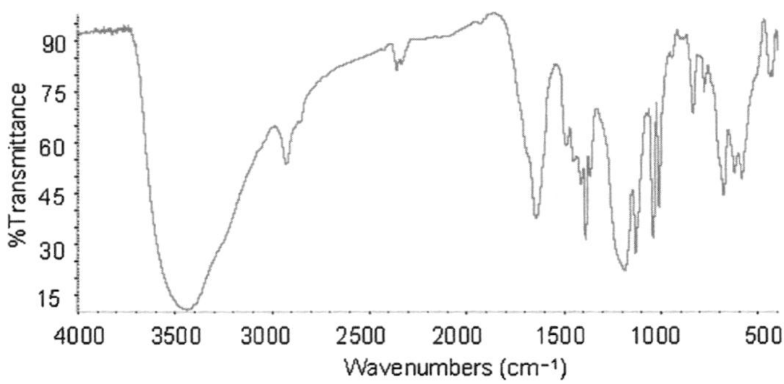

Fig. 6 FTIR spectra for enzymatic PEDOT/PSSNa film synthesized at an EDOT to PSSNa ratio of 1:1. (Reprinted with permission from Rumbau et al. [43]. © 2007, American Chemical Society)

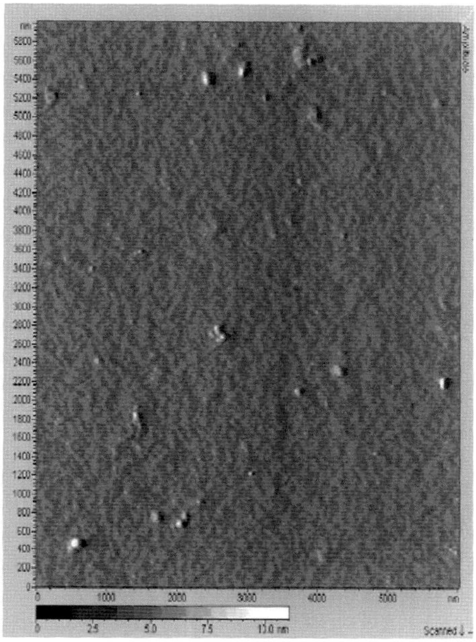

Fig. 7 Atomic force microscopy pictures obtained from a PEDOT/PSSNa film ($6\,\mu \times 6\,\mu$) synthesized at an EDOT to PSSNa ratio of 1:1. (Reprinted with permission from Rumbau et al. [43]. © 2007, American Chemical Society)

reaction mixture. The terthiophene monomer, having a slightly lower oxidation potential ($<1.02\,V$), acts as a redox mediator, and the polymerization of EDOT is performed in the presence of PSSNa (upper scheme, Scheme 5). This polymerization was observed not to proceed without the addition of terthiophene. When the polymerization of EDOT was performed under the same conditions without terthiophene, only low molecular weight oligomeric products were obtained [45]. In this work, as the concentration of terthiophene used was 100 times lower than that of EDOT monomer, the amount of oligomeric species formed by the oxidation of terthiophene was expected to be significantly low (1%). The oxidation of the EDOT monomer is the most difficult step in the polymerization process and the introduction of small amounts of terthiophene facilitated the process, as already established in earlier works on the enhancement of the electrochemical polymerization rate of thiophene by addition of terthiophene and/or bithiophene [46]. The authors extended the mechanism proposed by these earlier works to the polymerization of EDOT (lower scheme, Scheme 5), proposing that the terthiophene is incorporated along with the PEDOT in the final polymer.

The effect of concentration of the redox mediator on the formation of PEDOT/PSSNa was studied by these researchers by varying the concentration of terthiophene (0.025–1%) in a series of reactions. At least 0.5% (by weight of monomer concentration) of the terthiophene was observed to be required to initiate the polymerization of EDOT using SBP (Fig. 8). PEDOT/PSSNa was not formed when the concentration of terthiophene was 0.025%. The polymerization was also performed at various pH conditions, and the concentration of the final polymer was monitored spectroscopically, as shown in Fig. 8b. It could also be observed that the

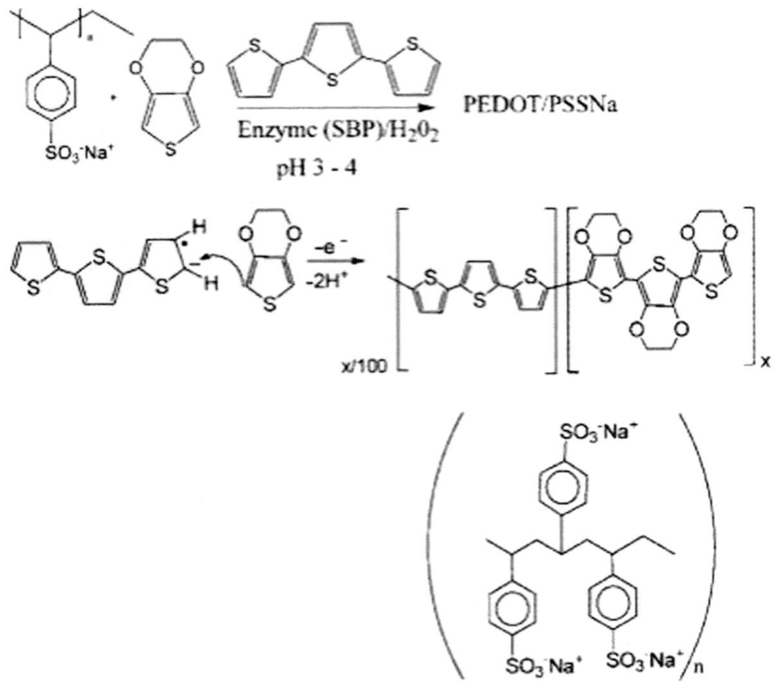

Scheme 5 *Above*: Polymerization of EDOT catalyzed by SBP using terthiophene as a redox mediator. *Below*: Proposed mechanism for the reaction. (Reprinted with permission from Nagarajan et al. [44]. © 2008, American Chemical Society)

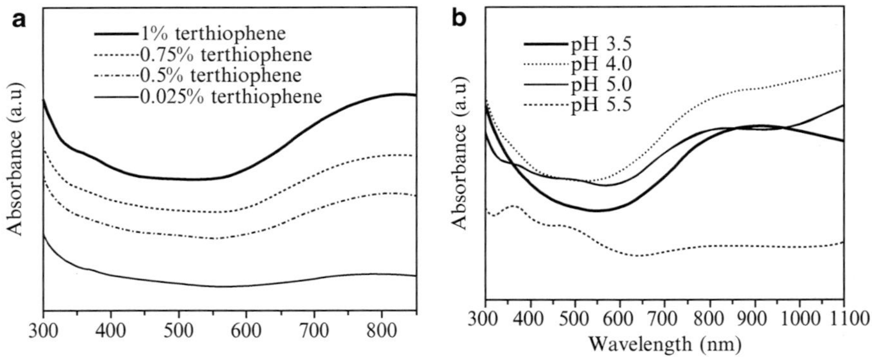

Fig. 8 (**a**) UV-Vis spectra for polymerization of EDOT at different terthiophene concentrations. (**b**) UV-Vis spectra for polymerization of EDOT at different pH values. (Reprinted with permission from Nagarajan et al. [44]. © 2008, American Chemical Society)

polymerization can be carried out even at pH 4.0 to yield doped PEDOT/PSSNa polymer, as seen by the absorption beyond 700 nm. The UV-Vis spectrum of the doped PEDOT/PSSNa closely resembled (Fig. 8) the PEDOT/PSSNa obtained from a commercial source (Aldrich).

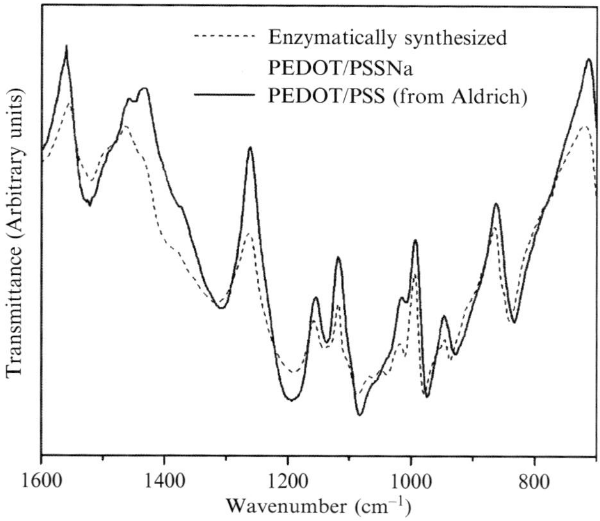

Fig. 9 IR spectra of commercial PDOT/PSSNa and enzymatically synthesized PEDOT/PSSNa. (Reprinted with permission from Nagarajan et al. [44]. © 2008, American Chemical Society)

The FTIR spectrum of the commercially available PEDOT/PSSNa and the enzymatically synthesized PEDOT/PSSNa were compared (Fig. 9). The spectra are scaled individually for clearer comparison. The vibrations at 1195, 1139, and 1089 cm^{-1} are due to the C–O–C bond stretch in the ethylenedioxy group. The peak at 1521 cm^{-1} is due to the ring stretching of the thiophene ring. The weak vibration at 1062 cm^{-1} is possibly due to the C–O stretch. Peaks at 979, 937, and 840 cm^{-1} are assigned to thiophene C–S bond stretching. As seen in Fig. 3, the PEDOT/PSSNa synthesized enzymatically shows similar features to those of the standard, and no major additional peaks are observed.

5 Biphasic Biocatalytic System for the Synthesis of Conducting Polymers

The use of enzymes as biocatalysts for the synthesis of water-soluble conducting polymers is simple, environmentally benign, and gives yields of over 90% due to the high efficiency of the enzyme catalyst. Since the use of an enzyme solution does not allow the recovery and reuse of the expensive enzyme, well-established strategies of enzyme immobilization onto solid supports have been applied to HRP [22–30]. A recent work reported an alternative method that allows the recycle and reuse of HRP in the biocatalytic synthesis of ICPs. The method is based on the use of a biphasic catalytic system in which the enzyme is encapsulated by simple solubilization into an IL. The main strategy consisted of encapsulating the HRP in room-temperature ILs insoluble in water, and the other components of the reaction (such as monomer, H_2O_2 oxidant, and dopant template) in the aqueous phase. The biocatalytic process should take place at the aqueous–IL interface, and the HRP is subsequently recovered by liquid–liquid phase separation after the biocatalytic reaction (Scheme 6).

This method was applied to the synthesis of both PANI [48] and PEDOT [47]. In both cases, after the polymerization reaction, the UV-Vis spectra of the obtained aqueous phase was compared to the corresponding conducting polymer obtained by classical biocatalytic synthesis, in which HRP is dissolved in the aqueous reaction media. In the case of PANI (Fig. 10), after the polymerization reaction, the UV-Vis

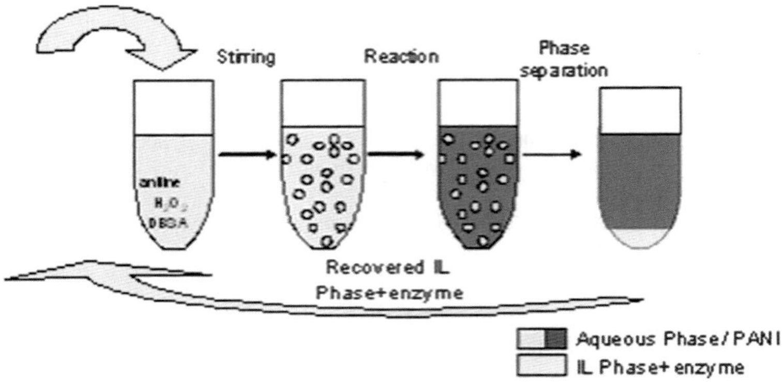

Scheme 6 Biphasic biocatalytic system in conducting polymer synthesis. (Reprinted with permission from Marcilla et al. [48]. © 2009, Wiley)

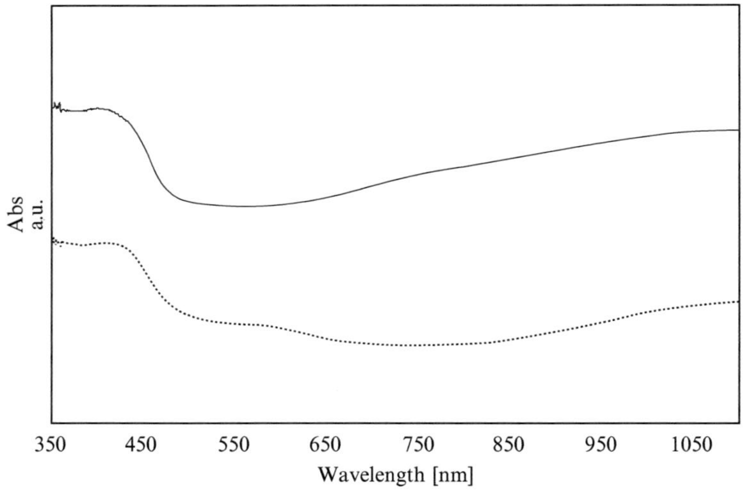

Fig. 10 UV-Vis spectra of PANI/DBSA aqueous solution obtained without encapsulation of the HRP enzyme (*upper spectrum*), and with encapsulation of the HRP enzyme in 1-butyl-3-methylimidazolium hexafluorophosphate (*lower spectrum*). (Reprinted with permission from Rumbau et al. [48]. © 2006, American Chemical Society)

spectra of the obtained aqueous phase was compared to the corresponding conducting polymer obtained by classical biocatalytic synthesis, in which HRP is dissolved in the aqueous reaction media. Both spectra were similar and clearly showed an absorption peak maximum at 420 nm. Furthermore, a well-known absorption tail extended towards the near-IR region between 800 and 1100 nm, arising from the polaron population in the material, which is a sign of the formation of conducting PANI. From the measurements obtained for PEDOT, the UV-Vis spectra of the EDOT monomer, the enzymatically obtained PEDOT, and a PEDOT prepared using the classical chemical oxidative polymerization [49, 50] are shown in Fig. 11. Both PEDOT spectra are similar, indicating a typical absorption peak maximum at 800 nm and proving the successful enzymatic synthesis of PEDOT.

The recovered HRP/IL mixture was added to a new aqueous reaction media, including fresh monomer, stabilizer, and H_2O_2. A second reaction took place and the HRP/IL was recovered again by liquid–liquid phase separation. In the case of aniline, the same process was successfully repeated up to five times using the same

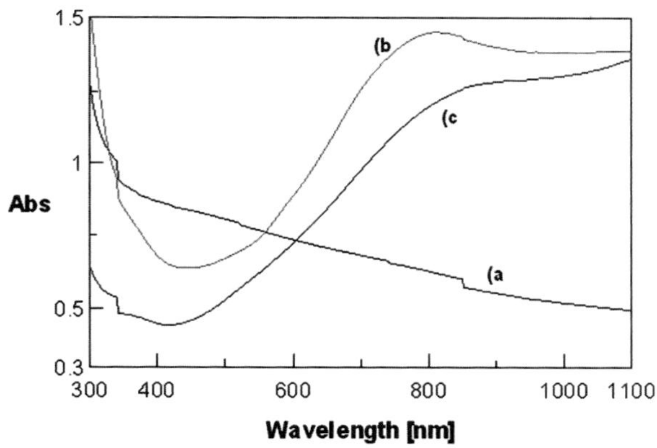

Fig. 11 UV-Vis spectra of (a) EDOT+PSS at pH 2, (b) enzimatically obtained PEDOT, and (c) PEDOT prepared using the classical chemical oxidative polymerization

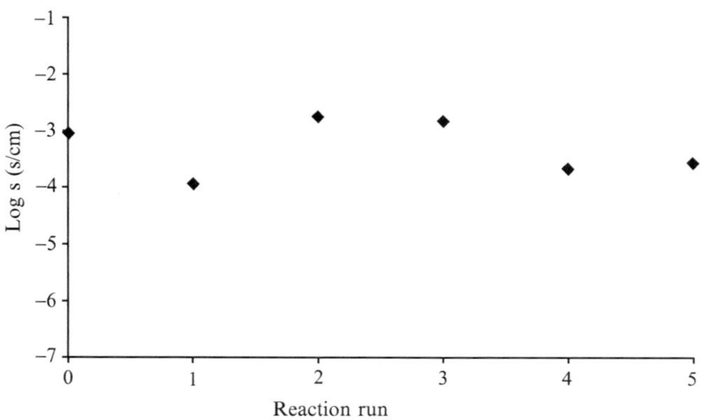

Fig. 12 Electrical conductivity films obtained from the aqueous solutions after several runs. *Run 0* indicates the control value obtained without encapsulation of the enzyme. (Reprinted with permission from Rumbau et al. [48]. © 2006, American Chemical Society)

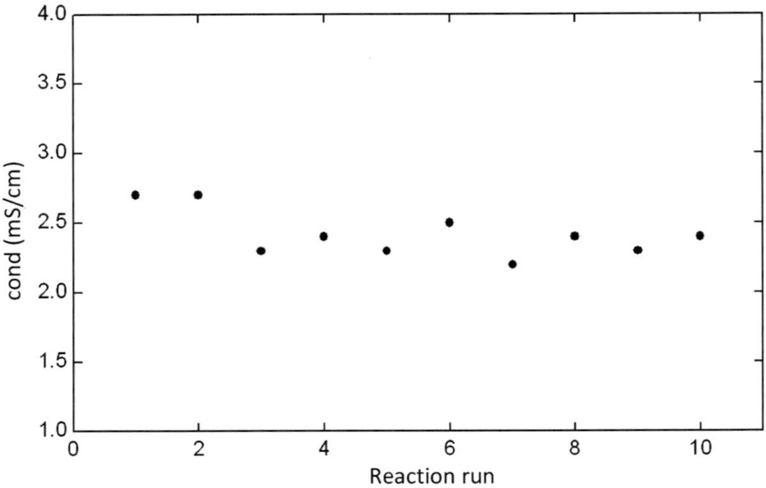

Fig. 13 Electrical conductivity of films obtained from the aqueous solutions after several runs. (Reprinted with permission from Sikora et al. [47]. © 2009, Wiley)

HRP/IL phase. The electrical conductivity of the PANI films prepared by solvent casting from the aqueous solutions showed a relatively high and similar value even after the fifth run (Fig. 12), which demonstrates the validity of our approach and the ease of recyclability and reuse of the enzyme inside the IL. For the PEDOT, the process of recovery and reuse was successfully repeated up to ten times using the same HRP/EDOT catalytic phase (Fig. 13), further confirming the success of the synthetic approach and the ease of recyclability and reuse of the enzyme inside the EDOT monomer phase.

6 Conclusions

This review chapter summarizes some of the most successful strategies recently introduced in the field of biocatalytic synthesis of ICPs as an attempt to solve the most problematic issues concerning the practical application of this technology. First, the use of a new bifunctional template in the enzymatic synthesis of PANI (DODD) is proposed, reaching, as a consequence, higher solubilities and electrical conductivities than those obtained when monofunctional templates are used. Second, a successfully performed enzymatic polymerization of EDOT in the presence of PSSNa is presented, proving that the biocatalytic strategy can also be applied to technologically interesting conducting polymers others than PANI. The obtained PEDOT polymer films showed excellent electrical conductivity (2×10^{-3} S cm^{-1}), as well as film formation capacity, with excellent surface roughness at the microscopic level. Finally, a new biphasic polymerization strategy is demonstrated that allows purification of the obtained conducting polymer aqueous solution from the remaining enzyme, so that the recovered enzyme can be recycled and reused several times.

Acknowledgments This work was supported by the Spanish Ministry of Science and Innovation under Contracts TEC2006–12376-C02, TEC2009–14446-C02–02, Project HOPE CSD2007–0007 (Consolider-Ingenio 2010) and Project HYPER CSD2009–00067 (Consolider-Ingenio 2010). Financial support from the Basque Government and the Diputacion de Gipuzkoa is gratefully acknowledged.

References

1. Nagarajan R, Tripathy S, Kumar J, Bruno FF, Samuelson L (2000) An enzymatically synthesized conducting molecular complex of polyaniline and poly(vinyl phosphonic acid). Macromolecules 33(26):9542–9547
2. Liu W, Kumar J, Tripathy S, Senecal KJ, Samuelson L (1999) Enzymatically synthesized conducting polyaniline. J Am Chem Soc 121(1):71–78
3. Annis BK, Narten AH, MacDiarmid AG, Richter AF (1988) A covalent bond to bromide in HBr-treated polyaniline from X-ray diffraction. Synth Met 22(3):191–199
4. Thiyagarajan M, Samuelson LA, Kumar J, Cholli AL (2003) Helical conformational specificity of enzymatically synthesized water-soluble conducting polyaniline nanocomposites. J Am Chem Soc 125(38):11502–11503
5. Wang X, Schreuder-Gibson H, Downey M, Tripathy S, Samuelson L (1999) Conductive fibers from enzymatically synthesized polyaniline. Synth Met 107(22):117–121
6. Liu W, Cholli AL, Kumar J, Tripathy S, Samuelson L (2001) Mechanistic study of the peroxidase-catalyzed polymerization of sulfonated phenol. Macromolecules 34(11): 3522–3526
7. Cruz-Silva R, Romero-García J, Angulo-Sanchez J, Ledezma-Perez J, Arias-Martín E, Moggio I, Flores-Loyola E (2005) Template-free enzymatic synthesis of electrically conducting polyaniline using soybean peroxidase. Eur Polym J 41(5):1129–1135
8. Xu P, Singh A, Kaplan DL (2006) Enzymatic catalysis in the synthesis of polyanilines and derivatives of polyanilines. Enzyme Catal Synth Polym 194:69–94
9. Gross RA, Kumar A, Kalra B (2001) Polymer synthesis by in vitro enzyme catalysis. Chem Rev 101:2097–2124
10. Kausaite A, Ramanaviciene A, Ramanavicius A (2009) Polyaniline synthesis catalysed by glucose oxidase. Polymer 50(8):1846–1851
11. Saunders BC, Holmes-Siedle AG, Stark BP (1964) Peroxidase: the properties and uses of a versatile enzyme and of some related catalysts. Butterworth, London
12. Alva KS, Kumar J, Marx KA, Tripathy SK (1997) Enzymatic synthesis and characterization of a novel water-soluble polyaniline: poly(2,5-diaminobenzenesulfonate). Macromolecules 30(14):4024–4029
13. Alva KS, Marx KA, Kumar J, Tripathy SK (1996) Biochemical synthesis of water soluble polyanilines: poly(*p*-aminobenzoic acid). Macromol Rapid Commun 17(12):859–863
14. Liu W, Wang JD, Ma L, Liu XH, Sun XD, Cheng YH et al (1995) Enzymatic polymerization of *p*-phenylphenol in aqueous micelles. Ann NY Acad Sci 750(1):138–145
15. Rao AM, John VT, Gonzalez RD, Akkara JA, Kaplan DL (1993) Catalytic and interfacial aspects of enzymatic polymer synthesis in reversed micellar systems. Biotechnol Bioeng 41(5):531–540
16. Premachandran RS, Banerjee S, John VT, McPherson GL, Akkara JA, Kaplan DL (1997) The enzymatic synthesis of thiol-containing polymers to prepare polymer-CdS nanocomposites. Chem Mater 9(6):1342–1347

17. Bruno R, Akkara JA, Samuelson LA, Kaplan DL, Marx KA, Kumar J et al (1995) Enzymatic mediated synthesis of conjugated polymers at the Langmuir trough air–water interface. Langmuir 11(3):889–892
18. Samuelson LA, Anagnostopoulos A, Alva KS, Kumar J, Tripathy SK (1998) Biologically derived conducting and water soluble polyaniline. Macromolecules 31(13):4376–4378
19. Liu W, Kumar J, Tripathy S, Senecal KJ, Samuelson LA (1999) Enzymatically synthesized conducting polyaniline. J Am Chem Soc 121(1):71–78
20. Guo ZW, Ruegger H, Kissner R, Ishikawa T, Willeke M, Walde P (2009) Vesicles as soft templates for the enzymatic polymerization of aniline. Langmuir 25(19):11390–11405
21. Streltsov AV, Shumakovich GP, Morozova OV, Gorbacheva MA, Yaropolov AI (2008) Micellar laccase-catalyzed synthesis of electroconductive polyaniline Appl Biochem Microbiol 44(3):264–270
22. Jim Z, Su Y, Duan Y (2001) A novel method for polyaniline synthesis with the immobilized horseradish peroxidase enzyme. Synth Met 122(2):237–242
23. Fernandes KF, Lima CS, Lopes FM, Collins CH (2004) Properties of horseradish peroxidase immobilised onto polyaniline. Proc Biochem 39(8):957–962
24. Azevedo AM, Vojinovic V, Cabral JMS, Gibson TD, Fonseca LP (2004) Operational stability of immobilised horseradish peroxidase in mini-packed bed bioreactors. J Mol Cat B Enzym 28(2–3):121–128
25. Sun D, Cai C, Li X, Xing W, Lu T (2004) Direct electrochemistry and bioelectrocatalysis of horseradish peroxidase immobilized on active carbon. J Electroanal Chem 566(2):415–421
26. Moeder M, Martin C, Koeller G (2004) Degradation of hydroxylated compounds using laccase and horseradish peroxidase immobilized on microporous polypropylene hollow fiber membranes. J Memb Sci 245(1):183–190
27. Jin Z, Su Y, Duan Y (2001) A novel method for polyaniline synthesis with the immobilized horseradish peroxidase enzyme. Synth Met 122:237–242
28. Oliveira GB, Filho JLL, Chaves MEC, Azevedo WM, Carvalho LB (2008) Enzyme immobilization on anodic aluminium oxide/polyethyleneimine or polyaniline composites. React Funct Polym 68(1):27–32
29. Alvarez S, Manolache S, Denes F (2003) Synthesis of polyaniline using horseradish peroxidase immobilized on plasma-functionalized polyethylene surfaces as initiator. J Appl Polym Sci 88(2):369–379
30. Yemini M, Xu P, Kaplan DL, Rishpon J (2006) Collagen-like peptide as a matrix for enzyme immobilization in electrochemical biosensors. Electroanalysis 18(21):2049–2054
31. Nabid MR, Golbabaee M, Moghaddam AB, Dinarvand R, Sedghi R (2008) Polyaniline/TiO_2 nanocomposite: enzymatic synthesis and electrochemical properties. Int J Electrochem Sci 3(10):1117–1126
32. Peng Y, Liu HW, Zhang XY, Li YS, Liu SY (2009) CNT templated regioselective enzymatic polymerization of phenol in water and modification of surface of MWNT thereby. J Polym Sci A Polym Chem 47(6):1627–1635
33. Bruno FF, Akkara JA, Kaplan DL, Sekher P, Marx KA, Tripathy SK (1995) Novel enzyme-mediated two-dimensional polymerization of aromatic derivatives on a Langmuir Trough. Ind Eng Chem Res 34(11):4009–4015
34. MacDiarmid AG, Epstein AJ (1997) Application of thin films of polyaniline and polypyrrole in novel light-emitting devices. ACS Symp Ser 672:395–415
35. Skotheim TA, Elsenbaumer RL, Reynolds JR (eds) (1998) Handbook of conducting polymers. Marcel Dekker, New York
36. Bruno FF, Nagarajan R, Roy S, Kumar J, Samuelson LA (2003) Biomimetic synthesis of water soluble conductive polypyrrole and poly(3,4-ethylenedioxythiophene). J Macromol Sci A Pure Appl Chem A 40(12):1327–1333
37. Bruno FF, Fossey SA, Nagarajan S, Nagarajan R, Kumar J, Samuelson LA (2006) Biomimetic synthesis of water-soluble conducting copolymers/homopolymers of pyrrole and 3,4-ethylenedioxythiophene. Biomacromolecules 7(2):586–589

38. Rumbau V, Pomposo JA, Alduncín JA, Grande H, Mecerreyes D, Ochoteco E (2007) A new bifunctional template for the enzymatic synthesis of conducting polyaniline. Enzyme Microb Technol 40(5):1412–1421
39. Dordick JS, Marletta MA, Klibanov AM (1987) Polymerization of phenols catalyzed by peroxidase in non-aqueous media. Biotechnol Bioeng 30:31–36
40. Akkara JA, Senecal KJ, Kaplan DL (1991) Synthesis and characterization of polymers produced by horseradish peroxidase in dioxane. J Polym Sci A Polym Chem 29:1561–1574
41. Ikeda R, Uyama H, Kobayashi S (1996) Laccase-catalyzed oxidative polymerization of syringic acid. Macromolecules 29:3053–3054
42. Ryu K, McEldoon JP, Pokora AR, Cyrus W, Dordick JS (1993) Numerical and Monte Carlo simulations of phenolic polymerizations catalyzed by peroxidase. Biotechnol Bioeng 42: 807–814
43. Rumbau V, Pomposo JA, Eleta A, Rodriguez J, Grande H, Mecerreyes D, Ochoteco E (2007) First enzymatic synthesis of water-soluble conducting poly(3,4-ethylenedioxythiophene). Biomacromolecules 8(2):315–317
44. Nagarajan S, Kumar J, bruno FF, Samuelson LA, Nagarajan R (2008) Biocatalytically synthesized poly(3,4-ethylenedioxythiophene). Macromolecules 41:3049–3052
45. Nagarajan R, Bruno FF, Samuelson LA, Kumar J (2004) Thiophene oligomer as a 'redox mediator' for the biocatalytic synthesis of poly(3,4-ethylenedioxythiophene) [PEDOT]. Polym Prepr 45:195–196
46. Wei Y, Chan CC, Tian J, Jang GW, Hsueh KF (1991) Electrochemical polymerization of thiophenes in the presence of bithiophene or terthiophene: kinetics and mechanisms of polymerization. Chem Mater 3:888–897
47. Sikora T, Marcilla R, Mecerreyes D, Rodriguez J, Pomposo JA, Ochoteco E (2009) Enzymatic synthesis of water-soluble conducting poly(3,4-ethylenedioxythiophene): a simple enzyme immobilization strategy for recycling and reusing. J Polym Sci A Polym Chem 47(1):306–309
48. Rumbau V, Marcilla R, Ochoteco E, Pomposo JA, Mecerreyes D (2006) Ionic liquid immobilized enzyme for biocatalytic synthesis of conducting polyaniline. Macromolecules 39(25):8547–8549
49. Heywang G, Jonas F (1992) Poly(alkylenedioxythiophene)s – new, very stable conducting polymers. Adv Mater 4(2):116–118
50. Kirchmeyer S, Reuter K (2005) Scientific importance, properties and growing applications of poly(3,4-ethylenedioxythiophene). J Mater Chem 15:2077–2088

Adv Polym Sci (2010) 237: 21–54
DOI:10.1007/12_2010_73

Published online: 28 May 2010

Transferases in Polymer Chemistry

Jeroen van der Vlist and Katja Loos

Abstract Transferases are enzymes that catalyze reactions in which a group is transferred from one compound to another. This makes these enzymes ideal catalysts for polymerization reactions. In nature, transferases are responsible for the synthesis of many important natural macromolecules. In synthetic polymer chemistry, various transferases are used to synthesize polymers in vitro. This chapter reviews some of these approaches, such as the enzymatic polymerization of polyesters, polysaccharides, and polyisoprene.

Keywords Biocatalysis · Enzymatic polymerization · Polyester · Polyisoprene · Polysaccharide · Transferase

Contents

J. van der Vlist and K. Loos (✉)
Department of Polymer Chemistry and Zernike Institute for Advanced Materials,
Faculty of Mathematics and Natural Sciences, University of Groningen,
Nijenborgh 4, 9747AG Groningen, The Netherlands
e-mail: K.U.Loos@rug.nl

Abbreviations

APP	Allylic diphosphate
Asp	Aspartic acid
ATRP	Atom transfer radical polymerization
CoA	Coenzyme A
CPT	*cis*-Prenyltransferase
Cys	Cysteine
DMAPP	Dimethylallyl diphosphate
DP	Degree of polymerization
EC	Enzyme Commision
FPP	Farnesyl diphosphate
GAG	Glycosaminoglycan
GGPP	Geranylgeranyl diphosphate
Glc	Glucose
GPP	Geranyl diphosphate
GT	Glycosyltransferase
HA	Hyaluronan
HAS	Hyaluronan synthase
His	Histidine
HMG-CoA	3-Hydroxy-methyl-glutaryl-CoA
IDI	Isopentenyl diphosphate isomerase
IDS	*trans*-Isoprenyl diphosphate synthase
IPI	IPP isomerase
IPP	Isopentenyl diphosphate
MPP-D	Mevalonate diphosphate decarboxylase
MW	Molecular weight
NADH	Nicotinamide adenine dinucleotide (reduced form)
NMR	Nuclear magnetic resonance
PHA	Polyhydroxyalkanoate
P3HB	Poly(3-hydroxybutyric acid)
TEMPO	2,2,6,6-Tetramethylpiperidine-1-oxyl
THF	Tetrahydrofuran
TPT	*trans*-Prenyltransferase
UDP	Uridine diphosphate

1 Introduction

Enzymatic polymerizations are an emerging research area with not only enormous scientific and technological promise, but also a tremendous impact on environmental issues. Biocatalytic synthetic pathways are very attractive as they have many advantages, such as mild reaction conditions, high enantio-, regio- and chemoselectivity, and the use of nontoxic natural catalysts.

Transferases (enzyme classification, class no. 2) are enzymes that catalyze reactions in which a group is transferred from one compound to another. Groups that are transferred are Cl, aldehydic or ketonic residues, acyl, glycosyl, alkyl, nitrogenous, and phosphorus- and sulfur-containing groups [1]. Of the three classes of enzymes used in polymer science so far, transferases are the least applied class of biocatalyst. Despite their potential for synthesizing interesting polymeric materials many transferases are very sensitive biocatalysts, which prevents their isolation on a larger scale and/or their use for synthesizing polymers on a reasonable scale.

The enzyme class of transferases is subdivided into nine subclasses:

EC 2 Transferases
EC 2.1 Transferring one-carbon groups
EC 2.2 Transferring aldehyde or ketonic groups
EC 2.3 Acyltransferases
EC 2.4 Glycosyltransferases
EC 2.5 Transferring alkyl or aryl groups, other than methyl groups
EC 2.6 Transferring nitrogenous groups
EC 2.7 Transferring phosphorus-containing groups
EC 2.8 Transferring sulfur-containing groups
EC 2.9 Transferring selenium-containing groups

This review focuses on acyl- and glycosyltransferases and transferases that transfer alkyl or aryl groups, other than methyl groups (EC 2.3, EC 2.4, and EC 2.5) as in these classes can be found interesting examples for the polymer scientist.

Acyltransferases are, for instance, able to synthesize biological polyesters with properties comparable or sometimes even exceeding polymers based on petrochemical-derived monomers. Acyltransferases are also frequently used to modify macromolecules in food and non-food applications.

For the synthesis of highly defined polysaccharides, glucosylsaccharides are the only option available. Specialized oligo- and polysaccharides for food and medical applications can be synthesized, and also hybrid structures with non-natural macromolecules or surfaces.

Prenyltransferases are responsible for the synthesis of *cis*-polyisoprene in natural rubber particles but can also be used to synthesize polyisoprenes in vitro.

2 Acyltransferases (EC 2.3)

2.1 Polyester Synthase

Polyhydroxyalkanoates (PHAs) are biological polyesters that are produced by a wide variety of bacteria as osmotically inert carbon- and energy-storage compounds that accumulate in the form of granules (see Fig. 1).

PHAs are generated in almost all bacteria under nutrient-limited growth conditions when a carbon source is readily available. PHA production occurs in exponential, late exponential, or in stationary growth phases, depending on the organism. Accumulation of PHA can reach as much as 85% of the dry cell weight. When the environment becomes more hospitable, the PHAs are degraded to the corresponding monomers, which are used as a source of energy for biosynthesis (supplying NADH) and as biosynthetic building blocks. As with all the other polymers discussed above, there are also conditions of growth in which PHA is generated transiently.

Polyhydroxyalkanoate synthase (systematic name: acyl-CoA:3-hydroxybutyrate O-acyltransferase; EC 2.3.1. class) is responsible for the polymerizations of PHAs in vivo because it catalyzes the stereoselective conversion of (*R*)-3-hydroxyacyl-CoA substrates to PHAs with the concomitant release of CoA (see Fig. 2) [3].

The residues Cys-319, Asp-480 and His-508 of the class I polyester synthase from *Cupriavidus necator* are conserved in all PHA synthases and were shown to be essential for covalent catalysis [2, 4, 5]. Cys-319 is the proposed catalytic nucleophile that is activated by the general base catalyst His-508.

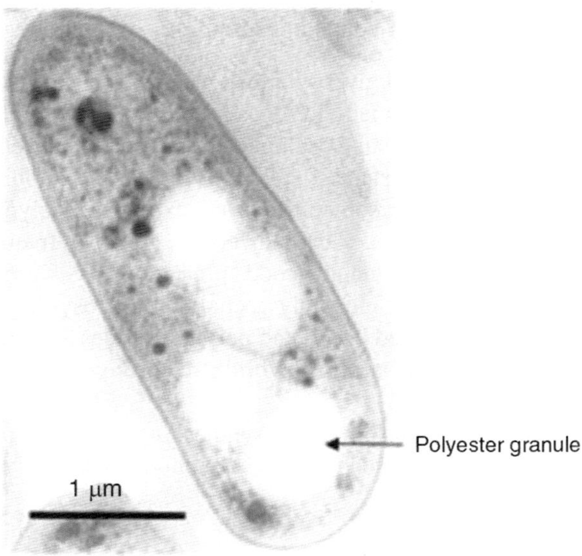

Fig. 1 Electron microscopy image of *Pseudomonas aeruginosa* harboring polyester granules [2] – Reproduced by permission of Portland Press Ltd.

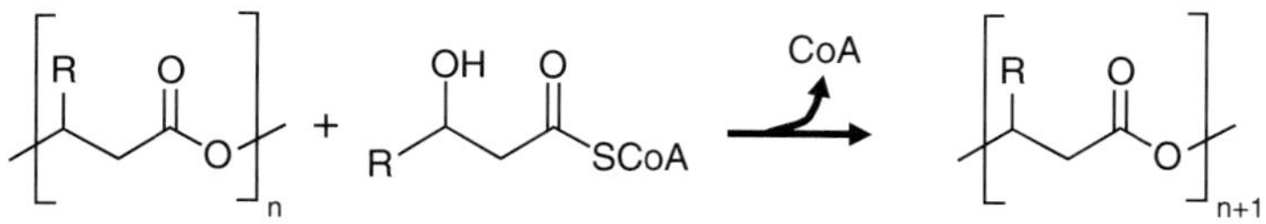

Fig. 2 Reaction catalyzed by polyester synthase

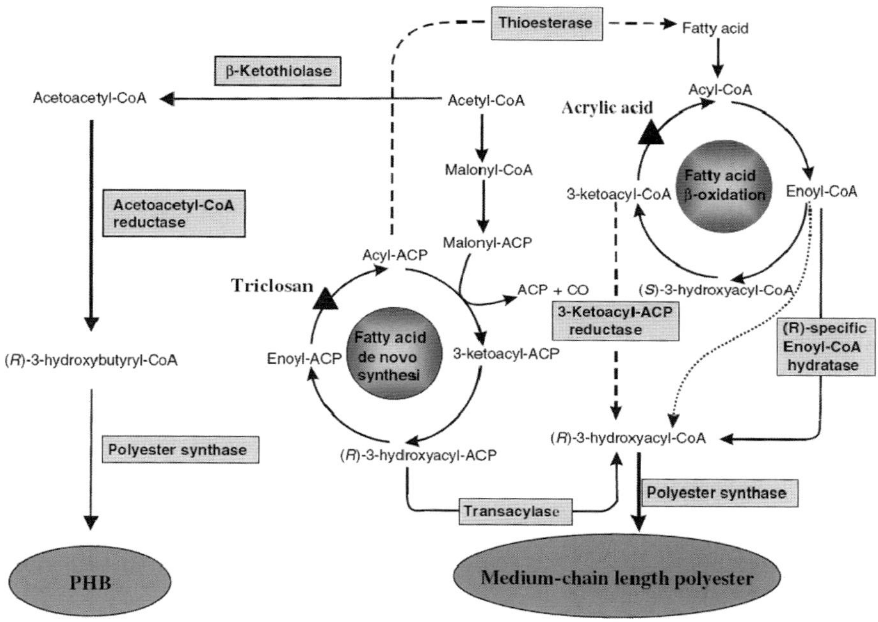

Fig. 3 Metabolic routes towards biopolyester synthesis. *Dashed lines* represent engineered biosynthesis routes. *Triangles* depict targets for inhibitors enabling biopolyester synthesis. Enzymes indicated on *shaded boxes on solid lines* are biopolyester biosynthesis enzymes. With kind permission from Springer Science+Business Media [7]

Reported efforts on engineering the PHA synthase towards better performance and selectivity in vivo and in vitro were recently assessed in an excellent mini-review by Nomura and Taguchi [6].

PHA polymers can be divided into three main classes with properties that are dependent on their monomer composition. Short-chain-length PHAs have monomers consisting of three to five carbons, are thermoplastic in nature, and generally lack toughness. Medium-chain-length PHAs have monomers consisting of 6–14 carbons and these polymers are elastomeric in nature. The third main class of PHAs includes copolymers made up of short-chain-length and medium-chain-length HA monomers consisting of 3–14 carbons. These PHAs have properties in between those of the first two classes, depending on the mole ratio of monomers, and therefore have a wide range of physical and thermal properties.

Figure 3 shows parts of the in vivo metabolic route towards PHAs. The biosynthesis of poly(3-hydroxybutyric acid) (P3HB) requires the condensation of two acetyl-CoA molecules catalyzed by the β-ketothiolase (PhaA), leading to the formation of acetoacetyl-CoA, which is reduced to (*R*)-3-hydroxybutyryl-CoA by the (*R*)-specific acetoacetyl-CoA reductase (PhaB). (*R*)-3-Hydroxybutyryl-CoA is substrate for the polyester synthase (PhaC) and the direct precursor of P3HB biosynthesis [2, 7]. In contrast to this, medium-chain-length PHAs are produced in vivo from intermediates of fatty acid metabolism (see Fig. 3).

PHAs can consist of a diverse set of repeating unit structures and have been studied intensely because the physical properties of these biopolyesters can be similar to petrochemical-derived plastics such as polypropylene (see Table 1). These biologically produced polyesters have already found application as bulk commodity plastics, fishing lines, and for medical use. PHAs have also attracted much attention as biodegradable polymers that can be produced from biorenewable resources. Many excellent reviews on the in vivo or in vitro synthesis of PHAs and their properties and applications exist, underlining the importance of this class of polymers [2, 6, 7, 12, 26–32].

In the large-scale production of PHAs, the extraction and purification of PHA from biomass is a crucial factor for determining the practical importance of these polymers. It is important that PHAs can be extracted efficiently and easily, much like the extraction of endogenous compounds such as starch, sucrose, and oil.

Table 1 Comparison of the physical properties of the poly(3-/4-hydroxyalkanoate)s with polypropylene. With kind permission from Springer Science+Business Media [8]

Polyester	T_g (°C)	T_m (°C)	Crystallinity	Elongation at break (%)	References
P3HB	15	175	50–80	5	[9]
P3HV	−15, 0	110, 112, 118	56		[10, 11]
P3HB-*co*-20mol%3HV	−1	145		50	[12]
P3HB-*co*-10mol%3HHx	−1	127			[13]
P3HB-*co*-17mol%3HHx	−2	120		850	[13]
P3HB-*co*-47mol% 3HV-*co*-16mol% 4HV-*co*-15mol%3HHx-*co*-2mol%3HO	−15	118		1,000	[11]
P3MB	8	100			[14]
P4HB	−40	53		1,000	[10]
P(3HB-*co*-16%4HB)			43	444	[15]
P3HPE	−11	63			[16, 17]
PHO	−36	59, 61			[18–20]
PHN	−39, −29	48, 54, 58			
P3H6MN	Not determined	65			[19, 20]
PH-p5TV	17	95			[21]
PH6PHx	4	Not determined			[22]
PH5PoxV	14	Not determined			[23]
PH-*p*-nitroPV	28	Not determined			[24]
PH8-pMPoxO	14	97			[25]
Polypropylene	−15	176	50	400	[12]

T_g glass transition temperature, T_m melting temperature

There are two common protocols used for PHA extraction from bacteria. The conventional one is based on the solubility of PHA in chloroform and insolubility in methanol [33]. After harvest, lipids and other lipophilic components in the bacterial cells are removed by reflux in hot methanol, followed by solubilization of PHA in warm chloroform. PHA from chloroform solvent can be recovered by solvent evaporation or precipitation by addition of methanol. Although highly purified PHA is obtained by this method, a large amount of hazardous solvent is needed to repeat the same process. Thus, this method is not environmentally friendly and unsuitable for mass production of bioplastic. The second protocol is designed to avoid the use of organic solvents. Bacterial cells are treated with a cocktail of enzymes (including proteases, nucleases, and lysozymes) and detergents to remove proteins, nucleic acids, and cell walls, leaving the PHA intact [34].

The first demonstration of in vitro P3HB synthesis in aqueous solution was achieved by Gerngross and Martin [35]. The polymer obtained had significantly higher molecular weight than that synthesized in vivo. This is probably due to a lack of a chain termination step of the PHA polymerization under in vitro conditions, similar to a living polymerization. Many other reports followed [27, 36, 38–44] and were extensively reviewed [2, 6, 7, 12, 26–32].

2.2 *Transglutaminase*

Transglutaminases (systematic name: protein-glutamine:amine gamma-glutamyltransferase; EC 2.3.2.13) belong to a class of enzymes known as amino-acyltransferases that catalyze calcium-dependent acyl transfer reactions between peptide-bound glutamine residues as acyl donors and peptide-bound lysine residues as acyl acceptors, resulting in the formation of intermolecular ε-(γ-glutamyl)lysine crosslinks.

The transamidation mechanism starts with the nucleophilic attack by the thiol group of an active-site cysteine residue (the catalytic triad is composed of Cys-276, His-334, and Asp-358) on the donor substrate γ-carboxamide group, leading to loss of an equivalent of ammonia and formation of a covalent thiolester intermediate. The acyl group of the transient thiolester is transferred to the acceptor amine substrate in the second step [45].

Transglutaminases form a family of related enzymes found in plasma, tissues, and extracellular media in all vertebrates. One of the best-known transglutaminases is the well-characterized human plasma coagulation enzyme Factor XIIIa, activated by thrombin from its tetrameric zymogen during the coagulation process. At a physiological level, tissue transglutaminases are involved in normal cellular processes, including cell adhesion [46], formation of the extracellular matrix [47], and apoptosis [48]. Diminished regulation of transglutaminase activity is implicated in a number of diseases.

Transglutaminases can recognize a broad range of primary amine acyl-acceptor substrates in vitro. However, recognition of the acyl-donor substrate in vivo is restricted to the γ-carboxyamide of glutamine within an apparently relatively small subset of sequence contexts

In vitro, the enzyme is able to catalyze crosslinking of whey proteins, soy proteins, wheat proteins, beef myosin, casein, and crude actomyosin (which is refined from mechanically deboned meat), improving functional properties such as the texture of food products [49–53]. Bonds formed by transglutaminase exhibit a high resistance to proteolytic degradation [54].

Although the main applications of microbial transglutaminases remain in the food sector, novel potential applications have emerged during the last decade. These applications cover the areas of biomedical engineering, material science, textiles and leather processing. For a recent excellent review by Zhu and Tramper see [55].

Transglutaminases can be used for improving the properties of protein-based fabrics such as wool, leading to a higher tensile strength after chemical or protease pretreatment [56, 57]. Besides crosslinking, transglutaminases were employed for grafting/coating of wool fabrics with silk sericin or keratin, leading to increased bursting strength and softness, and reduced felting shrinkage [58, 59].

In biomedical applications, transglutaminases have been used for tissue engineering materials such as enzymatically crosslinked collagen [60–63] or gelatin scaffolds [64–69]. Even melt-extruded guides based on enzymatically crosslinked macromolecules for peripheral nerve repair have been reported [70].

Transglutaminases have just become recently available in larger quantities and high purity due to microbiological production of the enzyme instead of extraction from animal tissue. It can be expected that this enzyme will be used more frequently in the future for food and non-food applications.

3 Glycosyltransferases (EC 2.4)

Glycosyltransferases (GTs) are important biological catalysts in cellular systems, generating complex cell surface glycans involved in adhesion and signaling processes. Recent advances in glycoscience have increased the demand to access significant amounts of glycans representing the glycome.

GTs catalyze the transfer of a sugar moiety from an activated donor sugar onto saccharide and nonsaccharide acceptors. GTs can be divided into the Leloir and non-Leloir types according to the type of glycosyl donors they use [71]. Non-Leloir glycosyltransferases typically use glycosyl phosphates as donors, whereas Leloir glycosyltransferases utilize sugar nucleotides as donors and transfer the monosaccharide with either retention (retaining enzymes) or inversion (inverting enzymes) of the configuration of the anomeric center. Most of the GTs responsible for the biosynthesis of mammalian glycoproteins and glycolipids are Leloir GTs.

GTs now play a key role for in vitro synthesis of oligosaccharides, and the bacterial genomes are increasingly utilized for cloning and overexpression of active transferases in glycosylation reactions [72–77]. In a recent excellent review by Homann and Seibel, possible ways to tailor-make biocatalysts by enzyme engineering and substrate engineering are summarized [78].

An established industrial process is the fermentative production of oligo- and polysaccharides (α-glucans) by lactic acid bacteria, e.g., dextran by *Leuconostoc mesenteroides*. Such polymers are used in the food industry, as additives for dyes, and in health care [79, 80]. Differences in the type of glycosidic linkage, the degree and type of branching, and the molecular mass of glucans available at present show promising variations in structural and functional properties, which need to be elucidated further. Known are dextrans with different structural specificities [α-1,6-bound glucose (Glc) backbone, with α-1,2 and α-1,3 side chains] [81, 82], mutan (α-1,3-bound Glc units) [83], alternan (α-1,6- and α-1,3-bound Glc units) [84], and amylose/reuteran (α-1,4- bound Glc units) [85]. In addition to defined polymers, oligosaccharides with specific structures are urgently needed. Oligosaccharides currently produced for commercial markets, including isomaltooligosaccharides, leucrose, and palatinose, are of interest in the fields of food, pharmacy, and cosmetics because of their ability to prevent and treat diseases from various biological origins [86, 87].

In this section, enzymes in the EC 2.4. class are presented that catalyze valuable and interesting reactions in the field of polymer chemistry. The Enzyme Commission (EC) classification scheme organizes enzymes according to their biochemical function in living systems. Enzymes can, however, also catalyze the reverse reaction, which is very often used in biocatalytic synthesis. Therefore, newer classification systems were developed based on the three-dimensional structure and function of the enzyme, the property of the enzyme, the biotransformation the enzyme catalyzes etc. [88–93]. The Carbohydrate-Active enZYmes Database (CAZy), which is currently the best database/classification system for carbohydrate-active enzymes uses an amino-acid-sequence-based classification and would classify some of the enzymes presented in the following as hydrolases rather than transferases (e.g. branching enzyme, sucrases, and amylomaltase) [91]. Nevertheless, we present these enzymes here because they are transferases according to the EC classification.

3.1 Phosphorylase

In the field of polymer science, the most extensively used transferase is phosphorylase (systematic name: (1→4)-α-D-glucan:phosphate α-D-glucosyltransferase; EC 2.4.1.1). Although this enzyme is responsible for the depolymerization of linear α-(1→4) glycosidic chains in vivo it can also be used to synthesize linear α-(1→4) glycosidic chains (amylose) in vitro.

In vivo linear α-1,4-glucans are synthesized from ADP-glucose by the enzyme glycogen synthase [94–97]. The enzyme, as well as the monomer, are quite sensitive and therefore most researchers (at least in the field of polymer science) prefer to use phosphorylase for the synthesis of amylose.

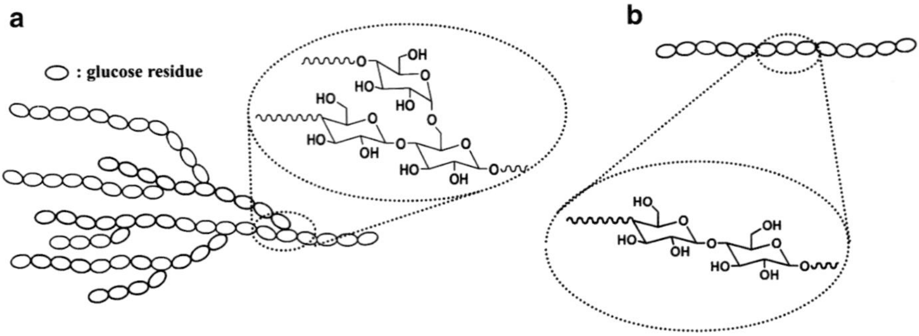

Fig. 4 Structure of (**a**) amylopectin and (**b**) amylose

Amylose is one component of starch, which is the most abundant carbohydrate storage reserve in plants. Carbohydrates such as starch function as a reservoir of energy for later metabolic use. It is found in many different plant organs, including seeds, fruits, tubers, and roots, where it is used as a source of energy during periods of dormancy and regrowth.

Starch granules are composed of two types of α-glucan, amylose and amylopectin, which represent approximately 98–99% of the dry weight. The ratio of the two polysaccharides varies according to the botanical origin of the starch.

Amylose is a linear molecule in which the glucose units are joined via α-(1→4) glucosyl linkages. Amylopectin is a branched molecule in which about 5% of the glucose units are joined by α-(1→6) glucosyl linkages (see Fig. 4).

In animals, a constant supply of glucose is essential for tissues such as the brain and red blood cells, which depend almost entirely on glucose as an energy source. The mobilization of glucose from carbohydrate storage provides a constant supply of glucose to all tissues. For this, glucose units are mobilized by their sequential removal from the non-reducing ends of starch. For this process three enzymes are required in vivo:

1. Glycogen phosphorylase catalyzes glycogen phosphorolysis (bond cleavage of the α-(1→4) bonds by the substitution of a phosphate group) to yield glucose-1-phosphate:

$$\text{glycogen (Gluc)}_\text{n} + P_i \leftrightarrows \text{glycogen (Gluc)}_{\text{n}-1} + \text{glucose} - 1 - \text{phosphate}$$

   Phosphorylase is only able to release glucose if the unit is at least five units away from a branching point.
2. Glycogen debranching enzyme removes α-(1→6) glycogen branches, thereby making additional glucose residues accessible to glycogen phosphorylase.
3. Phosphoglucomutase converts glucose-1-phosphate into glucose-6-phosphate, which has several metabolic fates.

The glycogen phosphorolysis of phosphorylase can be reverted, which makes it possible to enzymatically polymerize amylose as well as hybrid structures with amylose as outlined in the following section.

3.1.1 Enzymatic Polymerization of Amylose with Glycogen Phosphorylase

The existence of a phosphorylating enzyme in a higher plant was first reported by Iwanoff, who observed that an enzyme he found in the germinating vetches Vicia sativa liberates inorganic phosphate from organic phosphorous compounds [98]. Shortly after, the same enzyme was found in other vetches and wheat [99, 100], rice and coleseed [101], barley and malt etc. Bodńar was the first to report a progressive disappearance of inorganic phosphate (thus the reverse reaction) while incubating suspended flour from ground peas in a phosphate buffer [102]. Cori and Cori demonstrated that animal tissues contain an enzyme that acts upon glycogen as well [103–106]. Cori et al. suggested that the product of this reaction is α-glucopyranose-1-phosphoric acid (also called Cori-ester), which was confirmed later by Kiessling [107] and Wolfrom and Pletcher [108].

Glycogen phosphorylases belong to the group of vitamin B6 enzymes bearing a catalytic mechanism that involves the participation of the phosphate group of pyridoxal-5′-phosphate (PLP). The proposed mechanism is a concerted one with a front-side attack, as can be seen in Fig. 5 [109]. In the forward direction, e.g.,

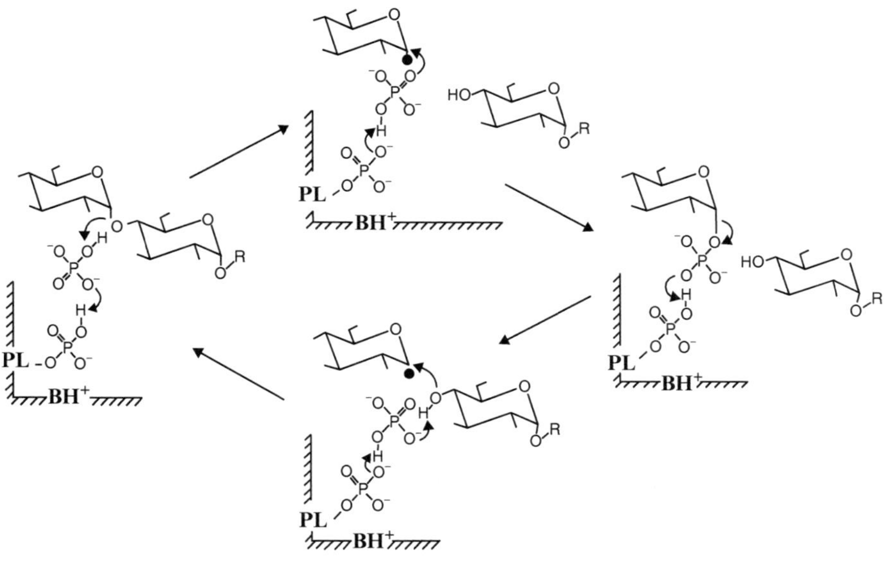

Fig. 5 Catalytic mechanism of glycogen phosphorylases. The reaction scheme accounts for the reversibility of phosphorolysis of oligosaccharides (*R*) in the presence of orthophosphate (*upper half*) and primer-dependent synthesis in the presence of glucose-1-phosphate (*lower half*). *PL* enzyme-bound pyridoxal; BH+ a general base contributed by the enzyme protein. Reprinted with permission from [109]. Copyright 1990 American Chemical Society

phosphorolysis of α-1,4-glycosidic bonds in oligo- or polysaccharides, the reaction is started by protonation of the glycosidic oxygen by orthophosphate, followed by stabilization of the incipient oxocarbonium ion by the phosphate anion, and subsequent covalent binding of the phosphate to form glucose-1-phosphate. The product, glucose-1-phosphate, dissociates and is replaced by a new incoming phosphate.

In the reverse direction, protonation of the phosphate of glucose-1-phosphate destabilizes the glycosidic bond and promotes formation of a glucosyl oxocarbonium ion–phosphate anion pair. In the subsequent step, the phosphate anion becomes essential for promotion of the nucleophilic attack of a terminal glucosyl residue on the carbonium ion. This sequence of reactions brings about α-1,4-glycosidic bond formation and primer elongation.

This mechanism accounts for retention of configuration in both directions without requiring sequential double inversion of configuration. It also provides for a plausible explanation of the essential role of PLP in glycogen phosphorylase catalysis: the phosphate of the cofactor PLP and the substrate phosphates approach each other within a hydrogen-bond distance, allowing proton transfer and making the phosphate of PLP into a proton shuttle that recharges the substrate phosphate anion.

The fact that glycogen phosphorylase can be used to polymerize amylose was first demonstrated by Schäffner and Specht [110] in 1938 using yeast phosphorylase. Shortly after, the same behavior was also observed for other phosphorylases from yeast by Kiessling [111, 112], muscles by Cori et al. [113], pea seeds [114] and potatoes by Hanes [115], and preparations from liver by Ostern and Holmes [116], Cori et al. [117] and Ostern et al. [118]. These results opened up the field of enzymatic polymerizations of amylose using glucose-1-phosphate as monomer, and can be considered the first experiments ever to synthesize biological macromolecules in vitro.

Recently Kuriki and coworkers succeeded in producing glucose-1-phosphate in situ during the enzymatic polymerization of amylose. By using sucrose phosphorylase or cellobiose phosphorylase, the monomer was produced during the polymerization from inorganic phosphate and sucrose or cellobiose, respectively [119–121].

One of the remarkable properties of phosphorylase is that it is unable to synthesize amylose unless a primer is added (poly- or oligomaltosaccharide):

$$\mathrm{n\,(glucose-1-phospate)+primer} \leftrightarrows \mathrm{amylose+n\,(orthophosphate)}$$

The kinetic behavior of the polymerization of amylose using potato phosphorylase with various saccharides as primers was first studied by Hanes [115]. Green and Stumpf [122] failed to detect priming action with maltose but were able to confirm all other results by Hanes. Weibull and Tiselius [123] found that the maltooligosaccharide of lowest molecular weight to exhibit priming activity was maltotriose, which was confirmed by Whelan and Bailey [124], who also showed that maltotriose is the lowest member of the series of oligosaccharides to exhibit priming activity (see Fig. 6).

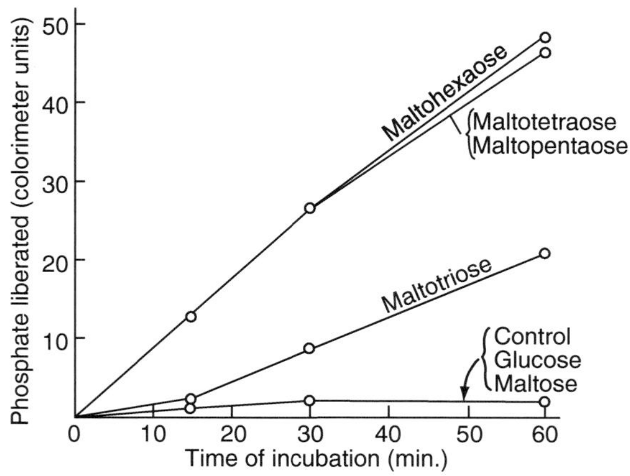

Fig. 6 Priming activity of glucose and maltooligosaccharides in the enzymatic polymerization using potato phosphorylase and glucose-1-phosphate as monomer [124] – Reproduced by permission of Portland Press Ltd.

Whelan and Bailey were also able to clarify the polymerization mechanism of the enzymatic polymerization with phosphorylase [124]. Their results showed that the polymerization follows a "multichain" scheme in contrast to a "single-chain" scheme that was also proposed by some authors. In the multichain polymerization scheme, the enzyme–substrate complex dissociates after every addition step, whereas in the single-chain scheme each enzyme continuously increases the length of a single primer chain without dissociation.

By studying the polydispersities of amyloses obtained by enzymatic polymerization with potato phosphorylase from maltooligosaccharides of various lengths, Pfannemüller and Burchard were able to show that the reaction mechanism of the polymerization with maltotriose as primer varies from its higher homologs [125]. The amyloses built by polymerization from maltotetraose or higher show a Poisson distribution [126], which can be expected from a polymerization following a multichain scheme (random synthesis occurs and all the primer chains grow at approximately equal rates). However, a bimodal broad distribution was observed when maltotriose was used as primer. The authors found that in the case of maltotriose as a primer the reaction can be divided into a start reaction and the following propagation, the rate of the first reaction being 400 times slower than the rate of the propagation. Due to this start reaction not all chains start to grow at the same time, which results in a broader distribution. The propagation again follows a multichain reaction scheme. Suganuma et al. [127] were able to determine the exact kinetic parameters of the synthetic as well as the phosphorolytic reaction using maltotriose and higher maltooligosaccharides as primer and were able to confirm the results of Whelan and Bailey [124] and Pfannemüller and Burchard [125].

3.1.2 Hybrid Structures with Amylose Blocks

The strict primer dependence of the glycogen phosphorylases makes them ideal candidates for the synthesis of hybrid structures of amylose with non-natural materials (e.g., inorganic particles and surfaces, synthetic polymers). For this, a primer functionality (maltooligosaccharide) can be coupled to a synthetic structure and subsequently elongated by enzymatic polymerization, resulting in amylose blocks. Various examples of these types of hybrid materials are outlined in the following section.

Amylose Hybrids with Short Alkyl Chains

Pfannemüller et al. showed that it is possible to obtain carbohydrate-containing amphiphiles with various alkyl chains via amide bond formation. For this, maltooligosaccharides were oxidized to the corresponding aldonic acid lactones, which could subsequently be coupled to alkylamines [128–136]. Such sugar-based surfactants are important industrial products with applications in cosmetics, medical applications etc. [137–139]. The authors were also able to extend the attached maltooligosaccharides by enzymatic polymerization using potato phosphorylase, which resulted in products with very interesting solution properties [140, 141].

Amylose Brushes on Inorganic Surfaces

Amylose brushes (a layer consisting of polymer chains dangling in a solvent with one end attached to a surface is frequently referred to as a polymer brush) on spherical and planar surfaces can have several advantageous uses, such as detoxification of surfaces etc. The modification of surfaces with thin polymer films is widely used to tailor surface properties such as wettability, biocompatibility, corrosion resistance, and friction [142–144]. The advantage of polymer brushes over other surface modification methods like self-assembled monolayers is their mechanical and chemical robustness, coupled with a high degree of synthetic flexibility towards the introduction of a variety of functional groups.

Commonly, brushes are prepared by grafting polymers to surfaces by, e.g., chemical bonding of reactive groups on the surface and reactive end groups of the attached polymers. This "grafting to" approach has several disadvantages as it is very difficult to achieve high grafting densities and/or thicker films due to steric crowding of reactive surface sites by already adsorbed polymers.

The "grafting from" approach (polymers are grown from initiators bound to surfaces) is a superior alternative because the functionality, density, and thickness of the polymer brushes can be controlled with almost molecular precision.

The first surface-initiated enzymatic polymerization reported was the synthesis of amylose brushes on planar and spherical surfaces [145]. For this, silica or silicone surfaces were modified with self-assembled monolayers of (3 aminopropyl)trimethoxysilane or chlorodimethylsilane, respectively. To these functionalities, oligosaccharides were added via (a) reductive amidation of the oligosaccharides to surface-bound amines, (b) conversion of the oligosaccharide to the according aldonic acid lactone and reaction with surface bound amines, and (c) incorporation of a double bond to the oligosaccharide and subsequent hydrosililation to surface-bound Si-H functions. The surface-bound oligosaccharides could be enzymatically elongated, using potato phosphorylase and glucose-1-phosphate as monomer, to amylose chains of any desired length. The degree of polymerization could be determined by spectrometric measurement of the liberated amount of inorganic phosphate [146], which was confirmed by cleavage of the amylose brushes (either enzymatically or by prior incorporation of light-sensitive spacers) and subsequent characterization of the free amylose chains. The obtained amylose-modified surfaces showed good chiral discrimination when employed as column materials in chiral affinity chromatography. Modification of the OH groups of the amylose brushes even enhanced the separation strength of the developed column materials (Loos, unpublished results). The results were recently confirmed by Breitinger, who attached maltooligosaccharides to surfaces via acid-labile hydrazide linkers, and enzymatically extended the chains using potato phosphorylase [147].

Copolymers with Amylose

The combination of oligo- or polysaccharides with non-natural polymeric structures opens up a novel class of materials. By varying the chain topology of the individual blocks as well as of the whole copolymer, the type of blocks, the composition etc., a complete set with tailor-made properties can be designed.

Amylose is a rod-like helical polymer consisting of α-(1$\rightarrow$4) glycosidic units. A measurement of the stiffness of a polymer is afforded by the so-called persistence length, which gives an estimate of the length scale over which the tangent vectors along the contour of the chains backbone are correlated. Typical values for persistence lengths in synthetic and biological systems can be several orders of magnitude larger than for flexible, coil-like polymers. Rod-like polymers have been found to exhibit lyotropic liquid crystalline ordered phases, such as nematic and/or layered smectic structures with the molecules arranged with their long axes nearly parallel to each other. Supramolecular assemblies of rod-like molecules are also capable of forming liquid crystalline phases. The main factor governing the geometry of supramolecular structures in the liquid crystalline phase is the anisotropic aggregation of the molecules.

Copolymeric systems with amylose are therefore systems in which at least one component is based on a conformationally rigid segment, and are generally referred to as rod–coil systems [148–151]. By combining rod-like and coil-like polymers, a novel class of self-assembling materials can be produced since the molecules share certain general characteristics typical of diblock molecules and thermotropic calamitic molecules. The difference in chain rigidity of rod-like and coil-like blocks is expected to greatly affect the details of molecular packing in the condensed phases and thus the nature of thermodynamically stable morphologies in these materials. The thermodynamically stable morphology probably originates as the result of the interdependence of microsegregation and liquid cristallinity. From this point of view, it is very fascinating to compare the microstructures originating in solution and in the bulk for such materials.

Practical applications in which copolymers are characterized by some degree of structural asymmetry have been suggested. For instance, a flexible block may be chosen because it donates a flexural compliance, whereas the more rigid portion offers tensile strength. In addition to the mechanical properties, the orientational order and the electrical conductance of certain rigid blocks could be exploited in optical and electrical devices.

Comb-type and linear block copolymer systems with enzymatically synthesized amylose have been reported and are outlined in the following section.

Comb-Type Copolymers with Amylose

The first comb-like structures synthesized by enzymatic "grafting from" polymerization from a polymeric backbone were reported by Husemann and Reinhardt [152, 153]. Acetobromo oligosaccharides were covalently bound to 6-trityl-2,3-dicarbanilyl-amylose chains and subsequently elongated by enzymatic polymerization using potato phosphorylase, the result being amylopectin-like structures with various degrees of branching. Pfannemüller and coworkers. extended this work by grafting amylose chains onto starch molecules. The modified starches were studied by the uptake of iodine and by light scattering measurements of carbanilate derivates [154] and appeared to be star-like in electron microscopy studies [155].

A full series of star-, network- and comb-like hybrid structures with oligosaccharides were synthesized by Pfannemüller and coworkers (see Fig. 7) and it was shown that the attached oligosaccharides can be extended via enzymatic polymerization using potato phosphorylase [128, 129, 136, 156, 157].

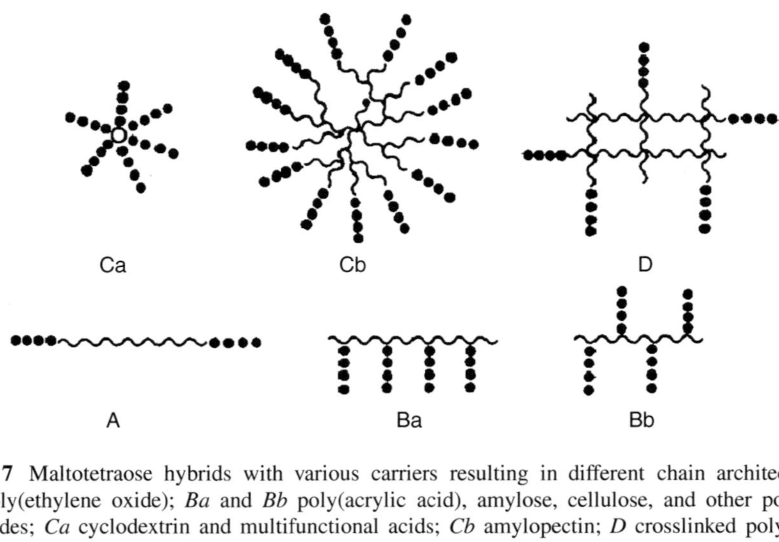

Fig. 7 Maltotetraose hybrids with various carriers resulting in different chain architectures: *A* poly(ethylene oxide); *Ba* and *Bb* poly(acrylic acid), amylose, cellulose, and other polysaccharides; *Ca* cyclodextrin and multifunctional acids; *Cb* amylopectin; *D* crosslinked poly(acryl amide) [156] – Reproduced by permission of Wiley

Another type of comb-like amylose hybrid synthesized via enzymatic grafting using phosphorylase is based on polysiloxane backbones. To achieve these structures, double bonds were incorporated to the reducing end of oligosaccharides, which were then attached to poly(dimethylsiloxane-*co*-methylsiloxane) copolymers via hydrosililation [158, 159] or to silane monomers, which were subsequently polymerized to polysiloxanes [160]. Various mono-, di-, tri- and oligosaccharides were attached to the siloxane backbones and their solution properties studied with viscosimetry and static and dynamic light scattering [161]. The pendant oligosaccharide moieties could be extended by enzymatic "grafting from" polymerization [162, 163].

Kobayashi and colleagues succeeded in attaching maltopentaose to the *para* position of styrene and performed free-radical polymerizations towards the homopolymers [164, 165] as well as towards copolymers with acrylamide [164]. Kobayashi and coworkers also reported the successful attachment of maltopentaose to poly(L-glutamic acid) [166]. Kakuchi and coworkers showed that the saccharide-modified styrene monomers could also be polymerized by TEMPO-mediated controlled radical polymerization [167]. Amylose-grafted polyacetylenes were recently reported by Kadokawa et al. [168]. Maltooligosaccharide-grafted polyacetylene was synthesized by Rh-catalyzed polymerizaton of *N*-propargylamide monomers having a maltooligosacchairde chain [168] and by attaching maltoheptaonolactone to amine-functionalized polyacetylene [169]. Kadokawa and coworkers also succeeded in covalently attaching maltooligosaccharides to natural biopolymers such as cellulose, chitin, and chitosan [170–172].

In all cases, the authors could successfully elongate the attached oligosaccharide structures using enzymatic polymerization, the product being comb-type block copolymers with amylose.

Linear Block Copolymers with Amylose

Various linear block copolymers of the AB, ABA, and ABC type containing enzymatically polymerized amylose blocks were reported. Ziegast and Pfannemüller converted the hydroxyl end groups of poly(ethylene oxide) into amino groups via tosylation and further reaction with 2-aminoalkylthiolate [173]. To the resulting mono- and diamino-functionalized poly(ethylene oxide), maltooligosaccharide lactones were attached and subsequently elongated to amylose via enzymatic polymerization [174]. Pfannemüller et al. performed a very detailed study on the solution properties of the synthesized ABA triblock copolymers because they can be considered model substances for "once-broken rod" chains [175]. With static and dynamic light scattering, the authors found that the flexible joint between the two rigid amylose blocks has no detectable effect on the common static and dynamic properties of the chain. Using dielectric measurements however, it became obvious that the directional properties of the electric dipoles of the broken rigid chains showed a different behavior to the non-broken rods (pure amylose). Akyoshi et al. also synthesized amylose-*b*-poly(ethylene glycol) block copolymers via enzymatic grafting from oligosaccharide-terminated poly(ethylene oxide) and studied the solution properties of these amphiphilic block copolymers by static and dynamic light scattering [176, 177].

It was also shown that the enzymatic polymerization of amylose could be started from oligosaccharide-modified polymers that are not soluble in the medium of polymerization (aqueous buffers). Amylose-*b*-polystyrene block copolymers could be synthesized by attaching maltooligosaccharides to anionically synthesized amino-terminated polystyrene, and subsequent enzymatic elongation to amylose [178, 179]. Block copolymers with a wide range of molecular weights and copolymer compositions were synthesized via this synthetic route. The solution properties of star-type as well as crew-cut micelles of these block copolymers were studied in water and THF, and the according scaling laws established [180]. In THF, up to four different micellar species were detectable, some of them in the size range of vesicular structures, whereas the crew-cut micelles in water were much more defined. Bosker et al. studied the interfacial behavior of amylose-*b*-polystyrene block copolymers at the air–water interface using the Langmuir–Blodgett technique [181].

Recently, two groups reported controlled radical polymerizations starting from maltooligosaccharides (ATRP [182] and TEMPO-mediated radical polymerization [183]), which will certainly lead to new synthetic routes towards amylose-containing block copolymers.

Even though the products are not block copolymer structures, the work of Kadokawa and colleagues should be mentioned here. In a process that the authors named "vine-twining polymerization" (after the way that a vine plant grow helically around a support rod), the enzymatic polymerization of amylose is performed in the presence of synthetic polymers in solution, and the authors showed that the grown amylose chains incorporate the polymers into its helical cavity while polymerizing [184–191].

3.2 *Branching Enzymes*

The formation of the α-(1→6) glucosyl branches of amylopectin and glycogen is synthesized by branching enzymes (systematic name: (1→4)-α-D-glucan:(1→4)-α-D-glucan 6-α-D-[(1→4)-α-D-glucano]-transferase; EC 2.4.1.18).

This enzyme catalyzes the formation of α-(1→6) branching points by cleaving an α-(1→4) glycosidic linkage in the donor substrate and transferring the nonreducing end-terminal fragment of the chain to the C-6 hydroxyl position of an internal glucose residue, which acts as the acceptor substrate [192]. Depending on its source, branching enzymes have a preference for transferring different lengths of glucan chains [193–196].

Recently, the in vitro synthesis of amylopectin- or glycogen-like structures via a tandem reaction of phosphorylase and branching enzyme was reported [197–201].

Phosphorylase catalyzes the polymerization of glucose-1-phosphate in order to obtain linear polysaccharide chains with α-(1→4) glycosidic linkages; the glycogen

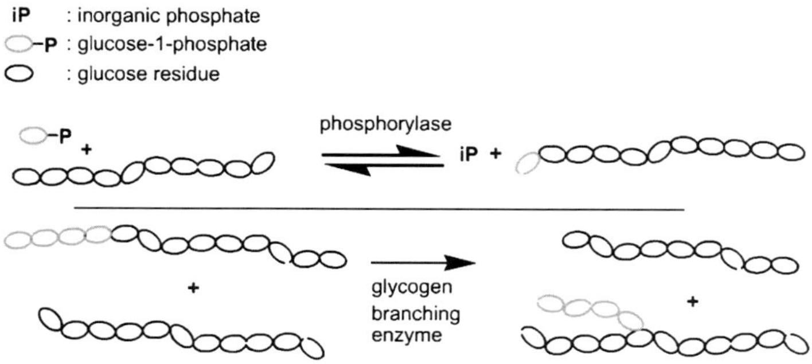

Fig. 8 Reactions catalyzed by glycogen phosphorylase (*above*) and glycogen branching enzyme (*below*)

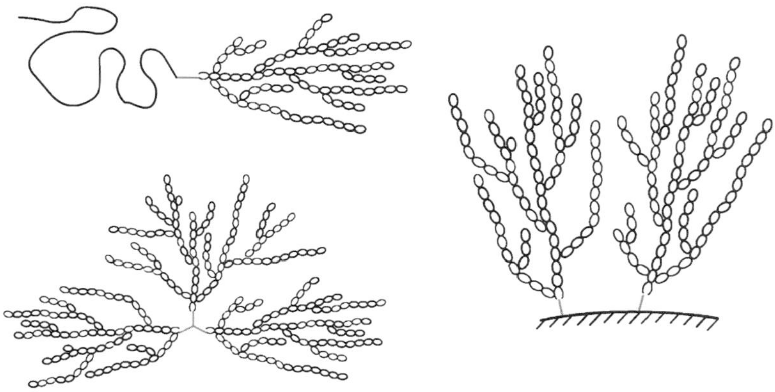

Fig. 9 Hybrid structures with amylopectin

branching enzyme is able to transfer short, α-(1→4) linked, oligosaccharides from the non-reducing end of starch to an α-(1→6) position (see Fig. 8). By combining the branching enzyme with phosphorylase, it becomes possible to synthesize branched structures via a one-pot synthesis because phosphorylase will polymerize linear amylose and the glycogen branching enzyme will introduce the branching points, which are again extended by phosphorylase.

As shown above, hybrid structures bearing amylose blocks can be synthesized by covalent attachment of primer recognition units for phosphorylase and subsequent enzymatic "grafting from" polymerization. Following the same route, we are currently synthesizing hybrid materials bearing (hyper)branched polysaccharide structures (as shown in Fig. 9) with the described tandem reaction of two enzymes. The branched structure, high amount of functional groups, and biocompatibility of these structures make these architectures suitable for applications in the biomedical field and in the food industry.

3.3 Glucansucrases

Another very valuable glycosyltransferase for the construction of well-defined polysaccharides is glucansucrase (dextransucrase; systematic name: sucrose:(1→6)-α-D-glucan 6-α-D-glucosyltransferase; EC 2.4.1.5).

Recently Dijkhuizen and coworkers reported on a family of glucansucrases found in *Lactobacillus reuteri*, which convert sucrose into large, heavily branched α-glucans [85]. One of these glucansucrases (GTF180) [202] produces an α-glucan with (α1 → 3) and (α1 → 6) glycosidic linkages. Kamerling and coworkers could show that the (1 → 3, 1 → 6)-α-D-glucan of *L. reuteri* strain 180 has a heterogeneous structure with no repeating units present [203]. It contains only α-D-Glcp-(1→6)-units in terminal position. All α-D-Glcp-(1→3) units were shown to be 6-substituted, and the polysaccharide is built-up from different lengths of isomalto-oligosaccharides, interconnected by single (α1 → 3) bridges. The unique polysaccharide structure produced was suggested to be prebiotic [204].

The GTF180 enzyme shows large similarity with other glucansucrase enzymes, but has a relatively large N-terminal variable region. Truncation of the enzyme, by deletion of the variable region, had no effect on the linkage distribution of the α-glucan produced [85].

Seibel and coworkers succeeded in constructing various new complex glycoconjugates containing thioglycosidic linkages to different glycopyranosides (galactose, glucose, neuraminic acid) (producing branched thiooligosaccharides) by changing the chemoselectivity of the various glucansucrases from α-1,6- to α-1,2-, α-1,3- or α-1,4-linked glucose [205].

Seibel and coworkers also showed that the mutagenesis is an effective tool for altering the regioselectivity and acceptor-substrate specificity of glucansucrase GTFR of *Streptococcus oralis*, a dextran-producing enzyme. By random mutagenesis, they were able to switch the regioselectivity and acceptor specificity of GTFR of *S. oralis* towards synthesis of (a) various short chain oligosaccharides or (b) novel (mutan) polymers with completely altered linkages, without compromising its high transglycosylation activity [206].

3.4 Levansucrase

Dijkhuizen and coworkers identified and characterized a Lactobacillus levansucrase (systematic name: sucrose:[6)-β-D-fructofuranosyl-(2→]n α-D-glucopyranoside 6-β-D-fructosyltransferase; EC 2.4.1.10) from *L. reuteri* strain 121, which could produce a high molecular weight levan polysaccharide from fructose [207].

3.5 Amylomaltase

Glycosyltransferases are also used extensively to modify natural polysaccharides. Thermoreversible gels that retrograded reversibly – comparable to gelatin gels – from enzymatically modified starch using amylomaltase were reported recently [208–213].

Amylomaltase (4-α-glucanotransferase; systematic name: (1→4)-α-D-glucan: (1→4)-α-D-glucan 4-α-D-glycosyltransferase; EC 2.4.1.25) catalyzes the glucan-chain transfer from one α-1,4-glucan to another α-1,4-glucan (or to the 4-hydroxyl group of glucose), or within a single linear glucan molecule to produce a cyclic-α-1,4-glucan [213–215]. The enzyme was first found in *Escherichia coli*, but seems to be distributed in various bacterial species with different physiological functions. In *E. coli*, amylomaltase is expressed with glucan phosphorylase from the same operon. Amylomaltase is a member of the maltooligosaccharide transport and utilization system and plays a role in converting short maltooligosaccharides into longer chains, upon which glucan phosphorylase can act [216]. The name amylomaltase is used for the microbial 4-α-glucanotransferases, whereas the plant counterparts are usually called disproportionating enzymes (D-enzymes) [212, 214, 215].

Typically, enzymatic modification of starch employs hydrolyzing enzymes such as α-amylase, pullulanase, and glucoamylase. These hydrolyze the α-1,4- or α-1,6-glycosidic bonds in amylose and amylopectin by first breaking the glycosidic linkage and subsequently using a water molecule as acceptor substrate. Amylomaltases also initially break the glycosidic linkage but, instead of water, they use another oligosaccharide as an acceptor substrate and form a new glycosidic linkage. Amylomaltases can use high molecular weight starch as both donor and acceptor molecule and can catalyze the transfer of long α-1,4-glucan chains [217], or even highly branched cluster units of amylopectin.

3.6 Hyaluronan Synthase

Hyaluronan (HA) is a nonsulfated non-epimerized linear glycosaminoglycan (GAG) existing in vivo as a polyanion of hyaluronic acid and composed of repeating disaccharide units of D-glucuronic acid and *N*-acetyl-D-glucosamine [GlcAβ(1→3)GlcNAcβ(1→4)] [218–220]. It is a major constituent of the extracellular matrix (ECM) of the skin, joints, eye, and many other tissues and organs. Despite the simple structure of this macromolecule, the complexity of its physicochemical properties and biological functions is tremendous. HA has extraordinary hydrophilic, rheological, and signaling properties and is viscoelastic. This naturally occurring biopolymer is dynamically involved in many biological processes, such as embryogenesis, inflammation, metastasis, tissue turnover, and wound healing.

The isolation, purification, and identification of nearly pure HA have been the center of scientific interest for many decades. The bacterial production of HA by *Streptococcus equi* [221] and *Streptococcus zooepidemicus* [222] enabled it to be produced in larger quantities than could be achieved by extraction methods alone. HA produced by *S. equi* has a lower molecular weight than does HA produced by *S. zooepidemicus*, which has a MW of about $1.8–2\times 10^6$ Da with a yield of around 4 g HA per liter of the cultivated solution. At present, HA from various sources, with different degrees of purity and molecular weights, is available for medical applications.

HA is synthesized by hyaluronan synthase (systematic name: alternating UDP-α-*N*-acetyl-D-glucosamine:β-D-glucuronosyl-(1→3)-[nascent hyaluronan] 4-*N*-acetyl-β-D-glucosaminyltransferase and UDP-α-D-glucuronate:*N*-acetyl-β-D-glucosaminyl-(1→4)-[nascent hyaluronan] 3-β-D-glucuronosyltransferase; EC 2.4.1.212) [223–226]. Hyaluronan synthase (HAS) is a single protein glycosyltransferase that is able to transfer two different monosaccharides, whereas most glycosyltransferases catalyze one glycosidic transfer reaction exclusively.

Markovitz et al. successfully characterized the HAS activity from *Streptococcus pyogenes* and discovered the enzyme's membrane localization and its requirements for sugar nucleotide precursors and Mg^{2+} [227]. DeAngelis et al. were the first to succeed in the molecular cloning and characterization of the Group A Streptococcal gene encoding the protein HasA, known to be in an operon required for bacterial HA synthesis [228, 229]. Following this, sequences of the genes encoding other HAS proteins were identified using molecular biological techniques [220, 224, 225, 230–243]. However, still little is known about the structure and mechanism of HAS.

The in vitro synthesis of HA oligomers and polymers using HAS and UDP–sugars was reported by the group of Paul DeAngelis. The monosaccharide units from UDP–GlcNAc and UDP–GlcA are transferred sequentially in an alternating fashion to produce the disaccharide repeats of the heteropolysaccharide. Recombinant derivatives of one HAS, PmHAS from the gram-negative bacterium *Pasteurella multocida* type A [233], have proved to be very useful for chemoenzymatic syntheses of both oligosaccharides [244] and polysaccharides [245, 246].

In 2004, the PmHAS was employed in synchronized, stoichiometrically controlled polymerization reactions in vitro to produce monodisperse HA polysaccharide preparations [246]. Reaction synchronization is achieved by providing the HAS with an oligosaccharide acceptor to bypass the slow polymer initiation step in vitro. All HA chains are elongated in parallel and thus reach the same length, yielding a population of narrow size distribution. The synthase adds all available UDP–sugar precursors to the nonreducing termini of acceptors in a non-processive fashion, as in the following equation:

$$\begin{aligned} &\mathrm{n(UDP-GlcA)+n(UDP-GlcNAc)+z(GlcA-GlcNAc)x} \rightarrow \mathrm{2n(UDP)} \\ &\quad \mathrm{+z(GlcA-GlcNAc)x+(n/z).} \end{aligned}$$

Therefore, size control is possible. For example, if there are many termini (i.e., z is large), then a limited amount of UDP–sugars will be distributed among many molecules and thus result in many short polymer chain extensions. Conversely, if there are few termini (i.e., z is small), then the limited amount of UDP–sugars will be distributed among few molecules and thus result in long polymer chain extensions. With this, it became possible to synthesize highly defined HA polymer standards that can be used for the characterization of polysaccharides by, for instance, size exclusion chromatography equipped with a multi-angle light scattering detector [247].

4 Transferases that Transfer Alkyl or Aryl Groups, Other than Methyl Groups (EC 2.5.)

Rubber is synthesized by plants via a side branch of the isoprenoid pathway by the enzyme rubber transferase (*cis*-prenyl transferase; systematic name: poly-*cis*-polyprenyl-diphosphate:isopentenyl-diphosphate polyprenylcistransferase; EC 2.5.1.20). Surprisingly, although this process has been studied for decades, due to the labile nature of the rubber transferase and the fact that it is a membrane-associated enzyme present in relatively low abundance, the identification of its protein subunits remain elusive. For some recent reviews on rubber biosynthesis, please refer to [248–251].

Natural rubber, *cis*-1,4-polyisoprene, is a strategically important plant-derived commodity required for the manufacture of numerous industrial, medical, and household items. Although more than 2500 plant species are known to produce rubber, *Hevea brasiliensis* (Brazilian rubber tree) is currently the sole source of natural rubber. Most countries depend on imports of *H. brasiliensis* rubber to sustain demand. Despite the increasing demand for natural rubber, the acreage for rubber trees has diminished in recent years. Furthermore, decades of inbreeding have rendered commercial *H. brasiliensis* varieties susceptible to abiotic stress and pathogen attack. Due to this, an increasing interest in the development of additional sources of natural rubber can be observed. Research on rubber-producing transgene plants, and on identification of the genes involved in its biosynthesis and regulation has intensified in recent years. To achieve a sustainable production of rubber from natural sources, our understanding of the molecular mechanism of rubber biosynthesis needs to be improved.

Prenyltransferases are a class of enzymes that transfer allylic prenyl groups to acceptor molecules. Prenyl transferases commonly refer to prenyl diphosphate synthases (even though the class of prenyl transferases also includes enzymes that catalyze the transfer of prenyl groups to acceptors that include not only isopentenyl diphosphate (IPP) but also aromatic compounds and proteins etc.).

The reactions catalyzed by prenyltransferases are unique and interesting from a mechanistic viewpoint. The reaction starts with elimination of the diphosphate ion from an allylic diphosphate (APP) to form an allylic cation, which is attacked by the IPP molecule, with stereospecific removal of a proton to form a new C–C bond and a new double bond in the product. By repeating this type of condensation between IPP and the allylic prenyl diphosphate product, prenyltransferase can synthesize a prenyl diphosphate with a certain length and stereochemistry fixed by its specificity (see Fig. 10) [252].

Rubber is synthesized and sequestered on cytsolic vesicles known as rubber particles. Rubber transferase is localized to the surface of the rubber particles, and biosynthesis is initiated through the binding of an allylic pyrophosphate (APP, a pyrophosphate, produced by soluble *trans*-prenyl transferases) primer. Progressive additions of IPP molecules ultimately result in the formation of high molecular weight *cis*-1,4-polyisoprene. The rubber transferase also requires a divalent cation, such as Mg^{2+} or Mn^{2+}, as cofactor.

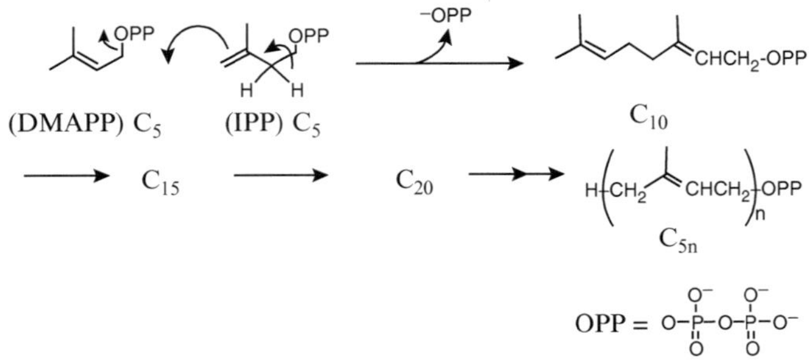

Fig. 10 Reactions catalyzed by prenyltransferases

The prenyl chain elongation catalyzed by prenyltransferases is quite unique because the reaction proceeds consecutively and terminates precisely at discrete chain lengths according to the specificities of the enzymes. The chain length of products varies so widely that it ranges from geraniol (C10) to natural rubber ($C > 5000$).

The identification and characterization of the genes and enzymes involved in rubber biosynthesis have been slow compared with those involved in the synthesis of other biopolymers. In fact, most of the studies thus far reported begin with rubber particles.

Enzymatically active, partially purified (washed) rubber particles can be isolated such that, when provided with an appropriate APP primer, magnesium ion cofactor, and IPP monomer, rubber is produced in vitro [253–255]. Fresh latex can be separated by centrifugation into three phases. The bottom fraction (20% of the latex) contains membrane-bound organelles. The middle fraction is called the C-serum. The top fraction phase contains the rubber particles. Biochemical studies have established that latex in this fractionated form is unstable. These studies also suggest that the bottom fraction is required for initiation of polymer synthesis.

The essential precursor (APP) and monomer (IPP) are synthesized in vivo via the isoprenoid pathway, with cytosolic acetyl-CoA being the primary building block for the synthesis of rubber. Acetyl-CoA is converted to IPP through a pathway involving the intermediate 3-hydroxy-methyl-glutaryl-CoA [256–258]. IPP is transformed into dimethylallyl diphosphate (DMAPP) by isomerization. DMAPP primes the sequential head-to-tail condensations of IPP molecules by *trans*-prenyltransferases to form geranyl diphosphate (GPP, C10, monoterpenoids), farnesyl diphosphate (FPP, C15, sesquiterpenoids) and geranylgeranyl diphosphate (GGPP, C20, diterpenoids) (Fig. 11).

Rubber molecules are synthesized from one APP molecule, which initiates the reaction, and the rubber polymer (*cis*-1,4-polyisoprene) is then polymerized by sequential condensations of the non-allylic IPP (magnesium cations are a required cofactor) with release of a diphosphate at each condensation. After initiation and elongation, a termination event occurs in which the rubber molecule is released from the enzyme. Despite the similar process, remarkable differences exist between plant species with respect to enzymatic reaction mechanisms and product molecular weight.

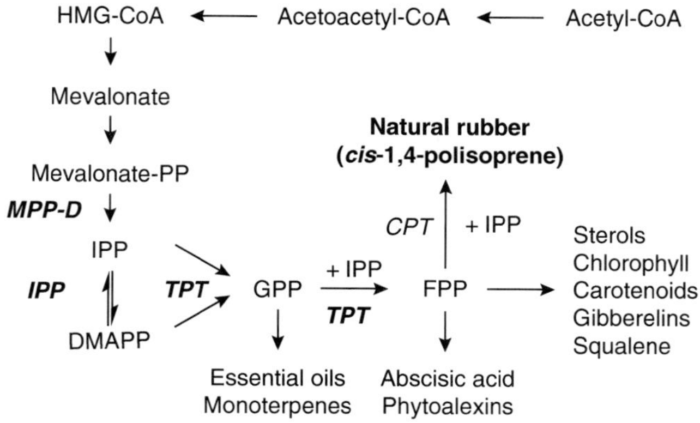

Fig. 11 Natural rubber is produced from a side branch of the ubiquitous isoprenoid pathway, with 3-hydroxy-methyl-glutaryl-CoA (HMG-CoA) as the key intermediate derived from acetyl-CoA by the general mevalonic-acid pathway. Mevalonate diphosphate decarboxylase (MPP-D) produces IPP, which is isomerized to DMAPP by IPP isomerase (IPI). IPP is then condensed in several steps with DMAPP to produce GPP, FPP and GGPP by the action of a *trans*-prenyltransferase (TPT). The *cis*-1,4-polymerization that yields natural rubber is catalyzed by *cis*-prenyltransferase (CPT), which uses the non-allylic IPP as substrate. Reprinted from [248], with permission from Elsevier

The exact mechanism of polymer initiation is unknown. Initiation of rubber synthesis has been studied in several plants and a common finding is that the end groups found in low molecular weight rubber (such as rubber from goldenrod and *H. brasiliensis* leaves) are not made up of *cis*-isoprene units, unlike the bulk of the rubber [259, 260]. Structural studies [261, 262] have led to the suggestion that the C15 FPP may be the most common initiator in vivo, at least in *H. brasiliensis*.

Although many different APPs are effective initiators of rubber biosynthesis, only IPP can be used as the source of isopentenyl monomer for the *cis*-1,4-polymerization of the rubber polymer.

All rubber transferases exhibit similar kinetic constants and pH optima, and are able to accept a similar range of APPs as initiating substrate [263, 264]. In vitro studies have shown that several compounds (DMAPP, GPP, FPP, and GGPP) can initiate rubber biosynthesis, with a faster rate of rubber biosynthesis the longer the APP (up to C15 or C20) [254, 265]. Non-natural APPs were also shown to be able to function as a primer for the rubber biosynthesis [266].

The cDNAs of the *cis*-prenyltransferase of *H. brasiliensis* was successfully identified and expressed in *E. coli*. The in vitro polymerization of IPP after initiation with FPP using the expressed *cis*-prenyltransferase resulted in low degrees of polymerization [267, 268]. After addition of rubber particles to this polymerization, the molecular weight increased tremendously [269]. It can be concluded that the rubber particles are essential for rubber biosynthesis. Katarina Cornish established a detailed structural model of the in vivo synthesis of natural rubber in the rubber particle monolayer membrane and partially explained this behavior (see Fig. 12) [251].

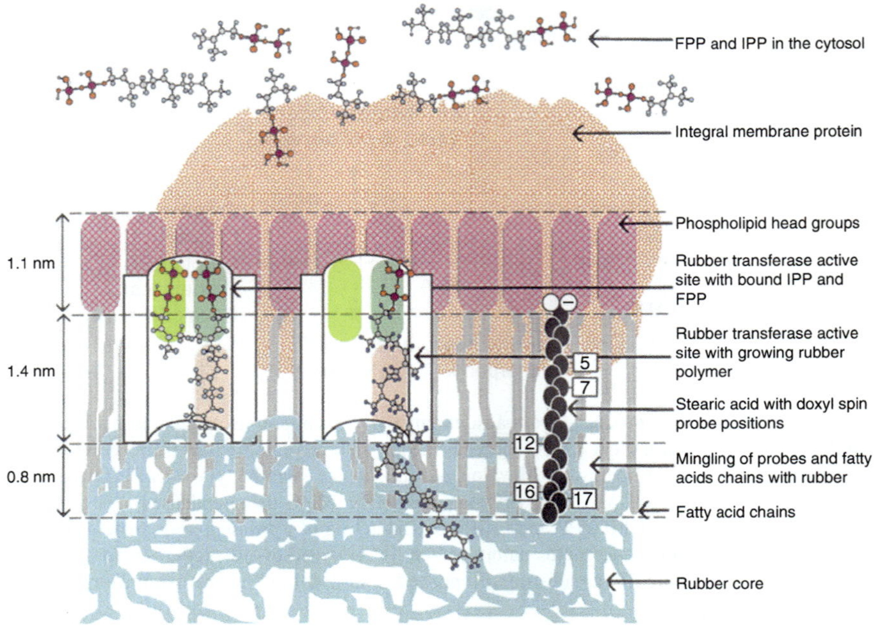

Fig. 12 Scheme depicting, to scale, a section of the surface monolayer biomembrane surrounding a *P. argentatum* rubber particle and the intermingling of the phospholipid fatty acid chains with the pure rubber core. Natural rubber is synthesized from IPP with an APP, most probably FPP, as an essential initiator. Both substrates are hydrophilic and are synthesized in the cytosol. The rubber transferase active site containing IPP (*yellow*) and allylic-PP (*green*) binding sites, and a non-specific hydrophobic binding region (*pink*) is drawn twice. The site on *left* of the schema contains bound IPP and FPP (the magnesium cofactor is not shown). Catalysis then would occur, leading to condensation of the two molecules and the release of pyrophosphate. The active site drawn in the *middle* of the schema is shown containing a short molecule of natural rubber (*cis*-1,4-polyisoprene) originally initiated by FPP. The *trans,trans* tail of the rubber molecule derived from FPP is visible. The active sites are positioned so that the hydrophilic pyrophosphate of FPP is aligned to the polar headgroup region of the membrane. Kinetic studies suggest that FPP traverses the active site as shown. Both models show the same physical position for the membrane–rubber interface [251] – Reproduced by permission of The Royal Society of Chemistry

Rubber appears to be a metabolic dead-end because there have been no findings of enzymes capable of breaking down the rubber in latex. The exact termination reaction of the rubber polymerization is not known. Different end-groups have been detected by NMR in rubber purified from a range of species, indicating that molecule dephosphorylation and release may involve esterification, cyclization, or hydrolysis [262].

In Hevea, the MW of the polymer has a bimodal distribution with some polymers in the range of 106 Da and others 105 Da. Both the Hevea polymers and the C55–C120 oligomeric isoprenes appear to have well-controlled molecular weights [270]. In vitro studies reveal the size of the polymer is related to the relative concentrations of the putative primer (farnesyl pyrophosphate or geranylgeranyl pyrophosphate)

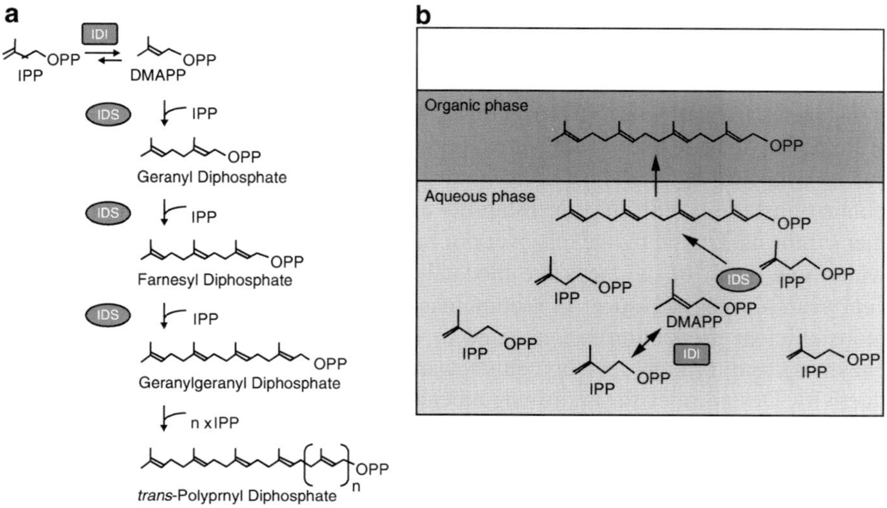

Fig. 13 Synthesis of polyisoprenyl diphosphates. (**a**) Biosynthetic pathway in archaea. (**b**) Concept of the organic–aqueous dual-phase system. The *light-* and *dark-shaded areas* indicate the aqueous and organic phases, respectively. *Double-headed and single-headed arrows* indicate isomerization by IDI and consecutive condensation by IDS, respectively. Reprinted from [271], with permission from Elsevier

and IPP [270]. The higher the ratio of primer to substrate, the shorter the chains. Thus, it is likely that the granule and its associated proteins in conjunction with the elongation protein(s) will play a critical role in chain length control and will be different from in vitro studies.

Recently Fujiwara et al. reported on the in vitro polymerization of *trans*-polyisoprene using the enzymes isopentenyl diphosphate isomerase (IDI) and *trans*-isoprenyl diphosphate synthase (IDS) [271]. IDI catalyzes the interconversion of IPP and DMAPP. IDS can now catalyze the polymerization of IPP from DMAPP as outlined above for the synthesis of natural rubber, and as outlined in Fig. 13a. However, the condensation process is inhibited due to hydrophobic interaction between IDS and hydrocarbon of the longer products. The hydrophobic chain of the elongating product does not readily protrude into the aqueous phase and it tends to interact with the enzyme. To achieve an efficient in vitro synthesis, the authors used an organic–aqueous two-liquid phase system to successfully synthesize (low molecular weight) *trans*-polyisoprene (see Fig. 13b).

5 Conclusions

It is obvious that transferases are powerful catalysts for the enzymatic synthesis of interesting polymer systems such as polyesters, polysaccharides, and polyisoprenes. Considering the range of reactions that transferases can in principle catalyze, it can be expected that more catalytic systems towards other polymeric materials will be developed in the future.

Despite their potential for synthesizing interesting polymeric materials many transferases are very sensitive biocatalysts, which prevents their isolation on a larger scale and/or their use for synthesizing polymer on a reasonable scale. However, it can be expected that with new biotechnological methods it will become possible to obtain transferases with enhanced stability that can be used more frequently in polymer synthesis. Driven by the discovery of many novel enzymes, synthesis routes in which one or all of the steps are biocatalytic have advanced dramatically in recent years through advances in recombinant DNA technology (which allows both more efficient production and targeted or combinatorial alterations of individual enzymes), and through developments towards higher stability and volumetric productivity. Design rules for improving biocatalysts are increasingly precise and easy to use.

Considering the macromolecules discussed in this chapter (polyester, polysaccharides, and rubber) there can be no doubt that new approaches to synthesize them via enzymatic polymerization will have a huge impact.

References

1. Nagel B, Dellweg H, Gierasch LM (1992) Pure Appl Chem 64:143
2. Rehm BHA (2003) Biochem J 376:15
3. Rehm BHA, Steinbuchel A (1999) Int J Biol Macromol 25:3
4. Jia Y, Yuan W, Wodzinska J, Park C, Sinskey AJ, Stubbe J (2001) Biochemistry 40:1011
5. Gerngross TU, Snell KD, Peoples OP, Sinskey AJ, Csuhai E, Masamune S, Stubbe J (1994) Biochemistry 33:9311
6. Nomura CT, Taguchi S (2007) Appl Microbiol Biotechnol 73:969
7. Rehm BHA (2006) Biotechnol Lett 28:207
8. Hazer B, Steinbuchel A (2007) Appl Microbiol Biotechnol 74:1
9. Doi Y (1990) Microbial polyesters. Wiley, New York
10. Choi MH, Yoon SC, Lenz RW (1999) Appl Environ Microbiol 65:1570
11. Schmack G, Gorenflo V, Steinbuchel A (1998) Macromolecules 31:644
12. Sudesh K, Abe H, Doi Y (2000) Prog Polym Sci 25:1503
13. Doi Y, Kitamura S, Abe H (1995) Macromolecules 28:4822
14. Kawada J, Lutke-Eversloh T, Steinbuchel A, Marchessault RH (2003) Biomacromolecules 4:1698
15. Doi Y, Segawa A, Kunioka M (1990) Int J Biol Macromol 12:106
16. Valentin HE, Berger PA, Gruys KJ, Rodrigues MFD, Steinbuchel A, Tran R, Asrar J (1999) Macromolecules 32:7389
17. Ulmer HW, Gross RA, Posada M, Weisbach P, Fuller RC, Lenz RW (1994) Macromolecules 27:1675
18. Gross RA, Demello C, Lenz RW, Brandl H, Fuller RC (1989) Macromolecules 22:1106
19. Hazer B, Lenz RW, Fuller RC (1994) Macromolecules 27:45
20. Hazer B, Erdem B, Lenz RW (1994) J Polym Sci Pol Chem 32:1739
21. Curley JM, Hazer B, Lenz RW, Fuller RC (1996) Macromolecules 29:1762
22. Hazer B, Lenz RW, Clinton Fuller R (1996) Polymer 37:5951
23. Ritter H, von Spee AG (1994) Macromol Chem Phys 195:1665
24. Arestogui SM, Aponte MA, Diaz E, Schröder E (1999) Macromolecules 32:2889

25. Kim YB, Kim DY, Rhee YH (1999) Macromolecules 32:6058
26. Suriyamongkol P, Weselake R, Narine S, Moloney M, Shah S (2007) Biotechnol Adv 25:148
27. Stubbe J, Tian JM, He AM, Sinskey AJ, Lawrence AG, Liu PH (2005) Annu Rev Biochem 74:433
28. Matsumura S (2002) Macromol Biosci 2:105
29. Steinbuchel A (2001) Macromol Biosci 1:1
30. van Beilen JB, Poirier Y (2008) Plant J 54:684
31. Lu JN, Tappel RC, Nomura CT (2009) Polymer Rev 49:226
32. Snell KD, Peoples OP (2009) Biofuels Bioprod Biorefining 3:456
33. Kessler B, Weusthuis R, Witholt B, Eggink G (2001) Adv Biochem Eng Biotechnol 71:159
34. Byrom D (1987) Trends Biotechnol 5:246
35. Gerngross TU, Martin DP (1995) Proc Natl Acad Sci USA 92:6279
36. Ren Q, de Roo G, van Beilen JB, Zinn M, Kessler B, Witholt B (2005) Appl Microbiol Biotechnol 69:286
37. Satoh Y, Murakami F, Tajima K, Munekata M (2005) J Biosci Bioeng 99:508
38. Tajima K, Satoh Y, Nakazawa K, Tannai H, Erata T, Munekata M, Kamachi M, Lenz RW (2004) Macromolecules 37:4544
39. Satoh Y, Tajima K, Tannai H, Munekata M (2003) J Biosci Bioeng 95:335
40. Kamachi M, Zhang SM, Goodwin S, Lenz RW (2001) Macromolecules 34:6889
41. Qi Q, Steinbuchel A, Rehm BHA (2000) Appl Microbiol Biotechnol 54:37
42. Su L, Lenz RW, Takagi Y, Zhang SM, Goodwin S, Zhong LH, Martin DP (2000) Macromolecules 33:229
43. Jossek R, Steinbuchel A (1998) FEMS Microbiol Lett 168:319
44. Jossek R, Reichelt R, Steinbuchel A (1998) Appl Microbiol Biotechnol 49:258
45. Folk JE (1969) J Biol Chem 244:3707
46. Gentile V, Thomazy V, Piacentini M, Fesus L, Davies PJA (1992) J Cell Biol 119:463
47. Aeschlimann D, Kaupp O, Paulsson M (1995) J Cell Biol 129:881
48. Oliverio S, Amendola A, DiSano F, Farrace MG, Fesus L, Nemes Z, Piredda L, Spinedi A, Piacentini M (1997) Mol Cell Biol 17:6040
49. Jaros D, Partschefeld C, Henle T, Rohm H (2006) J Texture Stud 37:113
50. De Jong GAH, Koppelman SJ (2002) J Food Sci 67:2798
51. Motoki M, Seguro K (1998) Trends Food Sci Technol 9:204
52. Zhu Y, Bol J, Rinzema A, Tramper J (1995) Appl Microbiol Biotechnol 44:277
53. Gauche C, Vieira JTC, Ogliari PJ, Bordignon-Luiz MT (2008) Process Biochem 43:788
54. Griffin M, Casadio R, Bergamini CM (2002) Biochem J 368:377
55. Zhu Y, Tramper J (2008) Trends Biotechnol 26:559
56. Cortez J, Bonner PLR, Griffin M (2005) J Biotechnol 116:379
57. Cortez J, Bonner PLR, Griffin M (2004) Enzyme Microb Technol 34:64
58. Cardamone JM (2007) Tex Res J 77:214
59. Cortez J, Anghieri A, Bonner PLR, Griffin M, Freddi G (2007) Enzyme Microb Technol 40:1698
60. Garcia Y, Wilkins B, Collighan RJ, Griffin M, Pandit A (2008) Biomaterials 29:857
61. O Halloran DM, Russell JC, Griffin M, Pandit AS (2006) Tissue Eng 12:1467
62. Chau DYS, Collighan RJ, Verderio EAM, Addy VL, Griffin M (2005) Biomaterials 26:6518
63. O Halloran DOGS, Stoddart M, Dockery P, Alini M, Pandit AS (2008) Biomaterials 29:438
64. Barbetta A, Massimi M, Di Rosario B, Nardecchia S, De Colli M, Devirgiliis LC, Dentini M (2008) Biomacromolecules 9:2844
65. Barbetta A, Massimi M, Devirgiliis LC, Dentini M (2006) Biomacromolecules 7:3059
66. Yung CW, Wu LQ, Tullman JA, Payne GF, Bentley WE, Barbari TA (2007) J Biomed Mater Res A 83A:1039
67. Broderick EP, O'Halloran DM, Rochev YA, Griffin M, Collighan RJ, Pandit AS (2005) J Biomed Mater Res B 72B:37
68. McDermott MK, Chen TH, Williams CM, Markley KM, Payne GF (2004) Biomacromolecules 5:1270
69. Lee KY, Mooney DJ (2001) Chem Rev 101:1869

70. Chiono V, Ciardelli G, Vozzi G, Cortez J, Barbani N, Gentile P, Giusti P (2008) Eng Life Sci 8:226
71. Leloir LF (1971) Science 172:1299
72. Lairson LL, Henrissat B, Davies GJ, Withers SG (2008) Annu Rev Biochem 77:521
73. Qian XP, Sujino K, Palcic MM, Ratcliffe RM (2002) J Carbohydr Chem 21:911
74. Davies GJ, Charnock SJ, Henrissat B (2001) Trends Glycosci Glycotechnol 13:105
75. Blixt O, Razi N (2008) In: Fraser-Reid BO, Tatsuta K, Thiem J (eds) Glycoscience: chemistry and chemical biology. Springer Berlin, Heidelberg, p. 1361
76. Song J, Zhang HC, Li L, Bi ZS, Chen M, Wang W, Yao QJ, Guo HJ, Tian M, Li HF, Yi W, Wang PG (2006) Curr Org Synth 3:159
77. Jakeman DL, Withers SG (2002) Trends Glycosci Glycotechnol 14:13
78. Homann A, Seibel J (2009) Appl Microbiol Biotechnol 83:209
79. Lacaze G, Wick M, Cappelle S (2007) Food Microbiol 24:155
80. Coviello T, Matricardi P, Alhaique F (2006) Expert Opin Drug DeliVery 3:395
81. Remaudsimeon M, Lopezmunguia A, Pelenc V, Paul F, Monsan P (1994) Appl Biochem Biotechnol 44:101
82. Robyt JF, Walseth TF (1979) Carbohydr Res 68:95
83. Kralj S, van Geel-Schutten GH, Dondorff MMG, Kirsanovs S, van der Maarel M, Dijkhuizen L (2004) Microbiology-(UK) 150:3681
84. Cote GL, Robyt JF (1982) Carbohydr Res 101:57
85. Kralj S, van Geel-Schutten GH, van der Maarel M, Dijkhuizen L (2004) Microbiology-(UK) 150:2099
86. Lina BAR, Jonker D, Kozianowski G (2002) Food Chem Toxicol 40:1375
87. Buchholz K, Noll-Borchers M, Schwengers D (1998) Starch-Starke 50:164
88. Kanehisa M, Goto S (2000) Nucleic Acids Res 28: 27
89. Kanehisa M, Goto S, Hattori M, Aoki-Kinoshita KF, Itoh M, Kawashima S, Katayama T, Araki M, Hirakawa M (2006) Nucleic Acids Res 34:D354
90. Greene LH, Lewis TE, Addou S, Cuff A, Dallman T, Dibley M, Redfern O, Pearl F, Nambudiry R, Reid A, Sillitoe I, Yeats C, Thornton JM, Orengo CA (2007) Nucleic Acids Res 35:D291
91. Cantarel BL, Coutinho PM, Rancurel C, Bernard T, Lombard V, Henrissat B (2009) Nucleic Acids Res 37:D233
92. Chang A, Scheer M, Grote A, Schomburg I, Schomburg D (2009) Nucleic Acids Res 37:D588
93. Kanehisa M, Goto S, Furumichi M, Tanabe M, Hirakawa M (2010) Nucleic Acids Res 38:D355
94. Ugalde JE, Parodi AJ, Ugalde RA (2003) Proc Natl Acad Sci USA 100:10659
95. Ball SG, Morell MK (2003) Annu Rev Plant Biol 54:207
96. Ball SG, van de Wal M, Visser RGF (1998) Trends Plant Sci 3:462
97. Guan HP, Keeling PL (1998) Trends Glycosci Glycotechnol 10:307
98. Iwanoff L (1902) Ber Deutsch Bot Ges 20:366
99. Zaleski W (1906) Ber Deutsch Bot Ges 24:285
100. Zaleski W (1911) Ber Deutsch Bot Ges 29:146
101. Suzuki U, Yoshimura K, Takaishi M (1906) Tokyo Kagaku Kaishi 27:1330
102. Bodńar J (1925) Biochem Z 165:1
103. Cori GT, Cori CF (1936) J Biol Chem 116:119
104. Cori GT, Cori CF (1936) J Biol Chem 116:129
105. Cori CF, Colowick SP, Cori GT (1937) J Biol Chem 121:465
106. Cori CF, Cori GT (1937) Proc Soc Exp Biol Med 36:119
107. Kiessling W (1938) Biochem Z 298:421
108. Wolfrom ML, Pletcher DE (1941) J Am Chem Soc; 63:1050
109. Palm D, Klein HW, Schinzel R, Buehner M, Helmreich EJM (1990) Biochemistry 29:1099
110. Schäfner A, Specht H (1938) Naturwissenschaften 26:494
111. Kiessling W (1939) Naturwissenschaften 27:129
112. Kiessling W (1939) Biochem Z 302:50
113. Cori CF, Schmidt G, Cori GT (1939) Science 89:464

114. Hanes CS (1940) Proc Roy Soc B 128:421
115. Hanes CS (1940) Proc Roy Soc B 129:174
116. Ostern P, Holmes E (1939) Nature 144:34
117. Cori GT, Cori CF, Schmidt G (1939) J Biol Chem 129:629
118. Ostern P, Herbert D, Holmes E (1939) Biochem J 33:1858
119. Ohdan K, Fujii K, Yanase M, Takaha T, Kuriki T (2007) J Biotechnol 127:496
120. Yanase M, Takaha T, Kuriki T (2006) J Sci Food Agric 86:1631
121. Ohdan K, Fujii K, Yanase M, Takaha T, Kuriki T (2006) Biocatal Biotransform 24:77
122. Green DE, Stumpf PK (1942) J Biol Chem 142:355
123. Weibull C, Tiselius A (1945) Arkiv för Kemi Mineralogi Och Geologi 19:1
124. Whelan WJ, Bailey JM (1954) Biochem J 58:560
125. Pfannemüller B, Burchard W (1969) Makromol Chem 121:1
126. Pfannemüller B (1975) Naturwissenschaften 62:231
127. Suganuma T, Kitazono JI, Yoshinaga K, Fujimoto S, Nagahama T (1991) Carbohydr Res 217:213
128. Emmerling WN, Pfannemüller B (1978) Makromol Chem 179:1627
129. Emmerling WN, Pfannemüller B (1981) Stärke 33:202
130. Pfannemüller B (1988) Stärke 40:476
131. Taravel FR, Pfannemüller B (1990) Makromol Chem 191:3097
132. Tuzov I, Cramer K, Pfannemüller B, Kreutz W, Magonov SN (1995) Adv Mater 7:656
133. Ziegast G, Pfannemüller B (1984) Makromol Chem 185:1855
134. Müller-Fahrnow A, Hilgenfeld R, Hesse H, Saenger W, Pfannemüller B (1988) Carbohydr Res 176:165
135. Pfannemüller B, Kühn I (1988) Makromol Chem 189:2433
136. Emmerling WN, Pfannemüller B (1983) Makromol Chem 184:1441
137. Biermann M, Schmid K, Schulz P (1993) Stärke 45:281
138. Hill K, Rhode O (1999) Fett-Lipid 101:25
139. von Rybinski W, Hill K (1998) Angew Chem-Int Edit 37:1328
140. Niemann C, Nuck R, Pfannemüller B, Saenger W (1990) Carbohydr Res 197:187
141. Ziegast G, Pfannemüller B (1987) Carbohydr Res 160:185
142. Zhao B, Brittain WJ (2000) Progr Polym Sci 25:677
143. Edmondson S, Osborne VL, Huck WTS (2004) Chem Soc Rev 33:14
144. Advincula R, Zhou QG, Park M, Wang SG, Mays J, Sakellariou G, Pispas S, Hadjichristidis N (2002) Langmuir 18:8672
145. Loos K, von Braunmühl V, Stadler R, Landfester K, Spiess HW (1997) Macromol Rapid Commun 18:927
146. Fiske CH, Subbarow Y (1925) J Biol Chem 66:375
147. Breitinger H-G (2002) Tetrahedron Lett 43:6127
148. Loos K, Munõz-Guerra S (2005) In: Ciferri A (ed) Supramolecular polymers. CRC, Boca Raton, p. 393
149. Stupp SI (1998) Curr Opin Colloid In 3:20
150. Klok HA, Lecommandoux S (2001) Adv Mater 13:1217
151. Lee M, Cho BK, Zin WC (2001) Chem Rev 101:3869
152. Husemann E, Reinhardt M (1962) Makromol Chem 57:109
153. Husemann E, Reinhardt M (1962) Makromol Chem 57:129
154. Burchard W, Kratz I, Pfannemüller B (1971) Makromol Chem 150:63
155. Bittiger H, Husemann E, Pfannemüller B (1971) Stärke 23:113
156. Andresz H, Richter GC, Pfannemüller B (1978) Makromol Chem 179:301
157. Emmerling W, Pfannemüller B (1978) Chem Ztg 102:233
158. Jonas G, Stadler R (1994) Acta Polymerica 45:14
159. Jonas G, Stadler R (1991) Makromol Chem-Rapid Commun 12:625
160. Haupt M, Knaus S, Rohr T, Gruber H (2000) J Macromol Sci Pure 37:323
161. Loos K, Jonas G, Stadler R (2001) Macromol Chem Phys 202:3210
162. von Braunmühl V, Stadler R (1996) Macromol Symp 103:141
163. von Braunmühl V, Jonas G, Stadler R (1995) Macromolecules 28:17

164. Kobayashi K, Kamiya S, Enomoto N (1996) Macromolecules 29:8670
165. Wataoka I, Urakawa H, Kobayashi K, Akaike T, Schmidt M, Kajiwara K (1999) Macromolecules 32:1816
166. Kamiya S, Kobayashi K (1998) Macromol Chem Phys 199:1589
167. Narumi A, Kawasaki K, Kaga H, Satoh T, Sugimoto N, Kakuchi T (2003) Polymer Bull 49:405
168. Kadokawa J, Nakamura Y, Sasaki Y, Kaneko Y, Nishikawa T (2008) Polymer Bull 60:57
169. Sasaki Y, Kaneko Y, Kadokawa J (2009) Polymer Bull 62:291
170. Omagari Y, Matsuda S, Kaneko Y, Kadokawa J (2009) Macromol Biosci 9:450
171. Kaneko Y, Matsuda S, Kadokawa J (2007) Biomacromolecules 8:3959
172. Matsuda S, Kaneko Y, Kadokawa J (2007) Macromol Rapid Commun 28:863
173. Ziegast G, Pfannemüller B (1984) Makromol Chem-Rapid Commun 5:363
174. Ziegast G, Pfannemüller B (1984) Makromol Chem-Rapid Commun 5:373
175. Pfannemüller B, Schmidt M, Ziegast G, Matsuo K (1984) Macromolecules 17:710
176. Akiyoshi K, Maruichi N, Kohara M, Kitamura S (2002) Biomacromolecules 3:280
177. Akiyoshi K, Kohara M, Ito K, Kitamura S, Sunamoto J (1999) Macromol Rapid Commun 20:112
178. Loos K, Müller AHE (2002) Biomacromolecules 3:368
179. Loos K, Stadler R (1997) Macromolecules 30:7641
180. Loos K, Böker A, Zettl H, Zhang AF, Krausch G, Müller AHE (2005) Macromolecules 38:873
181. Bosker WTE, Agoston K, Stuart MAC, Norde W, Timmermans JW, Slaghek TM (2003) Macromolecules 36:1982
182. Haddleton DM, Ohno K (2000) Biomacromolecules 1:152
183. Narumi A, Miura Y, Otsuka I, Yamane S, Kitajyo Y, Satoh T, Hirao A, Kaneko N, Kaga H, Kakuchi T (2006) J Polym Sci Pol Chem 44:4864
184. Kadokawa J-i, Kaneko Y, Nagase S, Takahashi T, Tagaya H (2002) Chemistry 8:3321
185. Kadokawa J-i, Kaneko Y, Nakaya A, Tagaya H (2001) Macromolecules 34:6536
186. Kadokawa J-i, Kaneko Y, Tagaya H, Chiba K (2001) Chem Commun 449
187. Kadokawa J-i, Nakaya A, Kaneko Y, Tagaya H (2003) Macromol Chem Phys 204:1451
188. Kaneko Y, Kadokawa J-i (2005) Chem Rec 5:36
189. Kaneko Y, Saito Y, Nakaya A, Kadokawa JI, Tagaya H (2008) Macromolecules 41:5665
190. Kaneko Y, Beppu K, Kadokawa JI (2008) Macromol Chem Phys 209:1037
191. Kaneko Y, Beppu K, Kadokawa JI (2007) Biomacromolecules 8:2983
192. Boyer C, Preiss J (1977) Biochemistry 16:3693
193. Palomo M, Kralj S, van der Maarel M, Dijkhuizen L (2009) Appl Environ Microbiol 75:1355
194. Abad MC, Binderup K, Rios-Steiner J, Arni RK, Preiss J, Geiger JH (2002) J Biol Chem 277:42164
195. Lerner LR, Krisman CR (1996) Cell Mol Biol 42:599
196. Takata H, Takata T, Okada S, Takagi M, Imanaka T (1996) J Bact 178:1600
197. van der Vlist J, Reixach MP, van der Maarel M, Dijkhuizen L, Schouten AJ, Loos K (2008) Macromol Rapid Commun 29:1293
198. Takata H, Kajiura H, Furuyashiki T, Kakutani R, Kuriki T (2009) Carbohydr Res 344:654
199. Fujii K, Takata H, Yanase M, Terada Y, Ohdan K, Takaha T, Okada S, Kuriki T (2003) Biocatal Biotransform 21:167
200. Kajiura H, Kakutani R, Akiyama T, Takata H, Kuriki T (2008) Biocatal Biotransform 26:133
201. Hernandez JM, Gaborieau M, Castignolles P, Gidley MJ, Myers AM, Gilbert RG (2008) Biomacromolecules 9:954
202. Pijning T, Vujicic-Zagar A, Kralj S, Eeuwema W, Dijkhuizen L, Dijkstra BW (2008) Biocatal Biotransform 26:12
203. van Leeuwen SS, Kralj S, van Geel-Schutten IH, Gerwig GJ, Dijkhuizen L, Kamerling JP (2008) Carbohydr Res 343:1237
204. Gibson GR, Roberfroid MB (1995) J Nutr 125:1401
205. Hellmuth H, Hillringhaus L, Hobbel S, Kralj S, Dijkhuizen L, Seibel J (2007) ChemBioChem 8:273

206. Hellmuth H, Wittrock S, Kralj S, Dijkhuizen L, Hofer B, Seibel J (2008) Biochemistry 47:6678
207. van Hijum S, Bonting K, van der Maarel M, Dijkhuizen L (2001) FEMS Microbiol Lett 205:323
208. Hansen MR, Blennow A, Pedersen S, Engelsen SB (2009) Carbohydr Polym 78:72
209. Hansen MR, Blennow A, Pedersen S, Norgaard L, Engelsen SB (2008) Food Hydrocolloids 22:1551
210. Lee KY, Kim YR, Park KH, Lee HG (2006) Carbohydr Polym 63:347
211. van der Maarel M, Capron I, Euverink GJW, Bos HT, Kaper T, Binnema DJ, Steeneken PAM (2005) Starch-Starke 57:465
212. Kaper T, Talik B, Ettema TJ, Bos H, van der Maarel M, Dijkhuizen L (2005) Appl Environ Microbiol 71:5098
213. Oh EJ, Choi SJ, Lee SJ, Kim CH, Moon TW (2008) J Food Sci 73:C158
214. Terada Y, Fujii K, Takaha T, Okada S (1999) Appl Environ Microbiol 65:910
215. Takaha T, Yanase M, Okada S, Smith SM (1993) J Biol Chem 268:1391
216. Takaha T, Smith SM (1999) Biotechnol Genet Eng Rev 16:257
217. Takaha T, Yanase M, Takata H, Okada S, Smith SM (1996) J Biol Chem 271:2902
218. Hargittai I, Hargittai M (2008) Struct Chem 19:697
219. Cowman MK, Matsuoka S (2005) Carbohydr Res 340:791
220. Lapcik L, Lapcik L, De Smedt S, Demeester J, Chabrecek P (1998) Chem Rev 98:2663
221. Mashimoto MS, H., Chiba S, Kitagawa H, Myoshi T (1988) Japan 63123392
222. Akasaka H, Susumu S, Yanagi M, Fukushima S, Mitsui T (1988) J Soc Cosmet Chem Jpn 22:35
223. Weigel PH, DeAngelis PL (2007) J Biol Chem 282:36777
224. Weigel PH (2002) IUBMB Life 54:201
225. Weigel PH, Hascall VC, Tammi M (1997) J Biol Chem 272:13997
226. DeAngelis PL (2002) Glycobiology 12:9R
227. Markovitz A, Cifonelli JA, Dorfman A (1959) J Biol Chem 234:2343
228. DeAngelis PL, Papaconstantinou J, Weigel PH (1993) J Biol Chem 268:19181
229. DeAngelis PL, Papaconstantinou J, Weigel PH (1993) J Biol Chem 268:14568
230. Blank LM, Hugenholtz P, Nielsen LK (2008) J Mol Evol 67:13
231. Jong A, Wu CH, Chen HM, Luo F, Kwon-Chung KJ, Chang YC, LaMunyon CW, Plaas A, Huang SH (2007) Eukaryot Cell 6:1486
232. DeAngelis PL, Jing W, Drake RR, Achyuthan AM (1998) J Biol Chem 273:8454
233. Spicer AP, McDonald JA (1998) J Biol Chem 273:1923
234. Kumari K, Weigel PH (1997) J Biol Chem 272:32539
235. DeAngelis PL, Jing W, Graves MV, Burbank DE, VanEtten JL (1997) Science 278:1800
236. Spicer AP, Olson JS, McDonald JA (1997) J Biol Chem 272:8957
237. DeAngelis PL, Achyuthan AM (1996) J Biol Chem 271:23657
238. Watanabe K, Yamaguchi Y (1996) J Biol Chem 271:22945
239. Shyjan AM, Heldin P, Butcher EC, Yoshino T, Briskin MJ (1996) J Biol Chem 271:23395
240. Spicer AP, Augustine ML, McDonald JA (1996) J Biol Chem 271:23400
241. Itano N, Kimata K (1996) Biochem Biophys Res Commun 222:816
242. Meyer MF, Kreil G (1996) Proc Natl Acad Sci U S A 93:4543
243. Itano N, Kimata K (1996) J Biol Chem 271:9875
244. DeAngelis PL, Oatman LC, Gay DF (2003) J Biol Chem 278:35199
245. Williams KJ, Halkes KM, Kamerling JP, DeAngelis PL (2006) J Biol Chem 281:5391
246. Jing W, DeAngelis PL (2004) J Biol Chem 279:42345
247. Jing W, Haller FM, Almond A, DeAngelis PL (2006) Anal Biochem 355:183
248. van Beilen JB, Poirier Y (2007) Trends Biotechnol 25:522
249. Puskas JE, Gautriaud E, Deffieux A, Kennedy JP (2006) Prog Polym Sci 31:533
250. Swiezewska E, Danikiewicz W (2005) Prog Lipid Res 44:235
251. Cornish K (2001) Nat Prod Rep 18:182
252. Ogura K, Koyama T (1998) Chem Rev 98:1263
253. Light DR, Dennis MS (1989) J Biol Chem 264:18589

254. Madhavan S, Greenblatt GA, Foster MA, Benedict CR (1989) Plant Physiol 89:506
255. Archer BL, Audley BG (1967) Adv Enzymol Relat Areas Mol Biol 29:221
256. Bouvier F, Rahier A, Camara B (2005) Prog Lipid Res 44:357
257. Sirinupong N, Suwanmanee P, Doolittle R, Suvachitanont W (2005) Planta 221:502
258. Fatland BL, Nikolau BJ, Wurtele ES (2005) Plant Cell 17:182
259. Cornish K (2001) Phytochemistry 57:1123
260. Ohya N, Koyama T (2001) In: Koyama T, Steinbüchel A (eds) Polyisoprenoids. Biopolymers, vol 2. Wiley, Germany, pp. 73–111
261. Tanaka Y (2001) Rubber Chem Technol 74:355
262. Tanaka Y (1989) Prog Polym Sci 14:339
263. Archer BL, Audley BG (1987) Bot J Linnean Soc 94:181
264. Cornish K, Siler DJ (1996) Plant Physiol Biochem 34:377
265. Cornish K, Siler DJ (1995) J Plant Physiol 147:301
266. Xie WS, McMahan CM, DeGraw AJ, Distefano MD, Cornish K, Whalen MC, Shintani DK (2008) Phytochemistry 69:2539
267. Ko JH, Chow KS, Han KH (2003) Plant MolBiol 53:479
268. Han KH, Shin DH, Yang J, Kim IJ, Oh SK, Chow KS (2000) Tree Physiol 20:503
269. Asawatreratanakul K, Zhang YW, Wititsuwannakul D, Wititsuwannakul R, Takahashi S, Rattanapittayaporn A, Koyama T (2003) Eur J Biochem 270:4671
270. Cornish K, Castillon J, Scott DJ (2000) Biomacromolecules 1:632
271. Fujiwara S, Yamanaka A, Yamada Y, Hirooka K, Higashibata H, Fukuda W, Nakayama JI, Imanaka T, Fukusaki EI (2008) Biochem Biophys Res Commun 365:118

Adv Polym Sci (2010) 237: 55–78
DOI:10.1007/12_2010_86

Published online: 10 June 2010

Hydrolases Part I: Enzyme Mechanism, Selectivity and Control in the Synthesis of Well-Defined Polymers

Martijn A.J. Veld and Anja R.A. Palmans

Abstract Lipases are highly activity in the polymerization of a range of monomers. Both ring-opening polymerization of cyclic esters and polycondensation reactions of AA–BB and AB monomers have been investigated in great detail. This paper reviews the increased understanding in enzymatic strategies for the production of well-defined polymers. Major advantages of enzymatic catalysts are the relatively mild reaction conditions, and the often-observed excellent regio-, chemo-, and enantioselectivity that allow for the direct preparation of functional materials. However, as a result of the monomer activation mechanism, polymers of low polydispersity and low quantitative degree of endgroup functionality are difficult to attain. A wide variety of (co)polymers have been synthesized and explored in a variety of applications using lipase catalysts.

Keywords (Co)polyesters · Lipase · Mechanism · Polycondensation · Ring-opening polymerization

Contents

M.A.J. Veld and A.R.A. Palmans (✉)
Laboratory of Molecular Science and Technology, Eindhoven University of Technology, PO Box 513, 5600 MB Eindhoven, The Netherlands
e-mail: m.a.j.Veld@tue.nl; a.palmans@tue.nl

Abbreviations

BHB	Bis(hydroxymethyl)butyric acid
CALB	*Candida antarctica* Lipase B
CL	ε-Caprolactone
DMP	2,4-Dimethyl-pentan-3-ol
DO	*p*-Dioxanone
DXO	1,5-Dioxepan-2-one
HEA	hydroxyethyl acrylate
HEMA	hydroxyethyl methacrylate
LCCC	Liquid chromatography under critical conditions
M_n	Number-average molecular weight
3MP	3-Mercaptopropionic acid
11MU	11-Mercaptoundecanoic acid
OC	2-Oxo-12-crown-4
pCL	Poly(ε-caprolactone)
PDL	ω-Pentadecanolactone
PDI	Polydispersity index
ROP	Ring-opening polymerization
SEC	Size exclusion chromatography
TMC	Trimethylene carbonate
VL	δ-Valerolactone

1 Introduction

Well-defined (co)polyesters and polyester-based architectures can nowadays be accessed by a wide variety of organometallic [1–4] and organocatalytic [5] catalysts, and applications in advanced materials for biomedical applications and nanotechnology are foreseen in the near future [6, 7]. Also, enzymes are perfectly equipped to convert monomers into polymers with high enantio-, regio-, and chemoselectivity [8, 9]. More specifically, in the last decade lipases have been shown to be outstanding catalysts for polymerization reactions as a result of their broad substrate scope, high activity, and excellent stability in a broad range of reaction media. Lipases are powerful catalysts for the preparation of polyesters, polycarbonates, and even polythioesters and polyamides. Moreover, a variety of different polymer architectures such as (block) copolymers and graft copolymers have been prepared using chemoenzymatic approaches [10].

This chapter will deal with the mechanistic features of lipases and how these affect the polymerizability of monomers. Although several lipases are capable of catalyzing polymerization reactions, Novozym 435 – an immobilized preparation of *Candida antarctica* Lipase B (CALB) on a crosslinked porous resin – is to date by far the most active biocatalyst. As a result, we will mainly focus on the use of CALB or Novozym 435 unless otherwise noted. Polyesters are especially interesting for future biomaterials and, as a result, approaches using lipase catalysis have been investigated in great detail over the past decade. The synthesis of other highly interesting polymers such as polycarbonates, polyamides, and polythioesters has been pursued with less rigor but a number of exciting results have appeared in literature. Excellent reviews have recently focused on the details of the different types of polymers that can be accessed via lipase catalysis [11–13]. In this chapter, we review the intrinsic possibilities of commercially available and easy-to-handle Novozym 435 in the field of enzymatic polymerizations, concentrating on results of the last decade. Moreover, we will also address the limitations that arise from applying lipases in the preparation of well-defined polymers.

2 Reaction Mechanism of Lipases and Implications for Monomer Acceptance in the Acylation and Deacylation Step

Lipases belong to the subclass of serine hydrolases, and their structure and reaction mechanism are well understood. Their common α/β-hydrolase enzyme fold is characterized by an α-helix that is connected with a sharp turn, referred to as the nucleophilic elbow, to the middle of a β-sheet array. All lipases possess an identical catalytic triad consisting of an Asp or Glu residue, a His and a nucleophilic Ser [14]. The latter residue is located at the nucleophilic elbow and is found in the middle of the highly conserved Gly−AA1−Ser−AA2−Gly sequence in which amino acids AA1 and AA2 can vary. The His residue is spatially located at one side of the Ser residue, whereas at the opposite side of the Ser a negative charge can be stabilized in the so-called oxyanion hole by a series of hydrogen bond interactions. The catalytic mechanism of the class of α/β-hydrolases is briefly discussed below using CALB as a typical example, since this is the most commonly applied lipase in polymerization reactions [15].

The catalytic triad of CALB consists of Asp187, His224, and Ser105 while the oxyanion hole is formed by the backbone amide protons of Thr40 and Gln106 and the side chain of Thr40 [16–18]. First, a substrate reversibly complexes to the free enzyme (Fig. 1 top left), thereby forming a Michaelis–Menten complex. After correct positioning of the substrate, nucleophilic attack of Ser105 onto the substrate carbonyl group occurs and a first tetrahedral intermediate is formed (Fig. 1 top right). In this tetrahedral intermediate, the negative charge on the former substrate carbonyl oxygen is stabilized by threefold hydrogen bonding interactions with the oxyanion hole, whereas the positive charge on His224 is stabilized by interaction

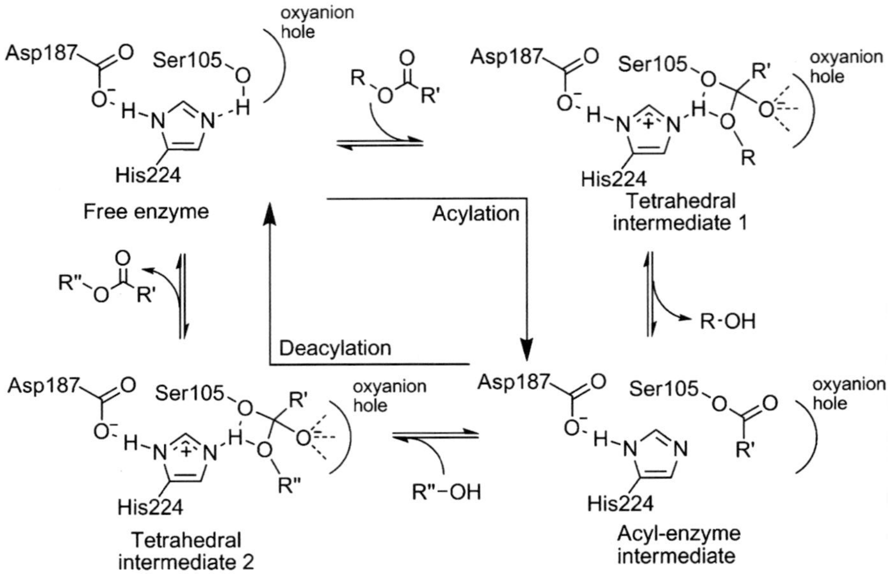

Fig. 1 Catalytic mechanism of CALB showing an acylation and deacylation step and the formation of a covalently bound acyl-enzyme intermediate (*bottom right*) [16]

with Asp187. Subsequently, proton transfer from His224 to the substrate alkyl oxygen takes place and the alcohol part of the substrate is liberated from the enzyme. As a result, a covalently bound acyl-enzyme intermediate is formed at the end of the acylation step (Fig. 1 bottom right). In case of polymerization reactions, this is referred to as the enzyme-activated monomer. Then, the acyl-enzyme intermediate is deacylated by an incoming nucleophile (R″NuH) which generally is water, an alcohol, or an amine. A second tetrahedral intermediate is formed by attack of the nucleophile onto the acyl-enzyme carbonyl group (Fig. 1 bottom left). In this process, the proton is transferred from the nucleophile to the His224 residue and the positive and negative charges are effectively stabilized by Asp187 and the oxyanion hole, respectively. Then, the proton is transferred from the His224 residue to the Ser105 alkyl oxygen while restoring the carbonyl bond of the bound substrate. As a result, a weakly bound enzyme–product complex is formed and the free enzyme species is regenerated after release of the reaction product.

The natural substrates of lipases are triglycerides and, in an aqueous environment, lipases catalyze their hydrolysis into fatty acids and glycerol. In anhydrous media, lipases can be active in the reverse reaction [19]. In fact, in the acylation step, acids, lactones, (cyclic) carbonates [20, 21], cyclic amides [22, 23], (cyclic) thioesters [24, 25], and cyclic phosphates [26] have been found to act as suitable acyl donors. In the deacylation step, apart from water, lipases also accept alcohols [27], amines [28, 29], and thiols [30] as nucleophiles although the specificity of lipases is lower for amines and thiols than for water and alcohols [31].

Table 1 Overview of nucleophiles and acyl donors accepted by lipases and the resulting polymers

(Di)acyl donor	(Di)nucleophile	Polymers prepared	References
Polycondensation			
R-CO-OH	Alcohol	Polyesters	[32–41]
	Thiol	Polythioesters	[42, 43]
R-CO-OR′	Alcohol	Polyesters	[44–51]
	Amine	Polyamides	[52–54]
	Thiol	Polythioesters	[25, 43]
R-O-CO-OR′	Alcohol	Polycarbonates	[55–57]
Ring-opening polymerization			
R-CO-OR′ (cyclic)	Alcohol/water	Polyesters	[58–81]
R-O-CO-OR′ (cyclic)	Alcohol/water	Polycarbonates	[82–93]
R-CO-NH-R′ (cyclic)	Water	Polyamides	[22]
R-CO-SR′ (cyclic)	Water	Polythioesters	[25]

Combining (di)acyl donors with (di)nucleophiles allows access to a wide variety of functional groups and, thus, polymers. Table 1 gives an overview of the different polymers that have been prepared employing lipase catalysis. Two types of polymerization reactions are distinguished: the polycondensations and the ring-opening polymerizations (ROPs) of cyclic monomers. The latter require only a small amount of nucleophile to initiate the polymerization reaction, whereas in the polycondensations, acyl donor and nucleophile must be present in stoichiometric amounts. The following sections will deal in detail with the different types of polymers – polyesters, polycarbonates, polyamides, and polythioesters – that have been prepared using lipase catalysis.

3 Polyesters

The preparation of a variety of (co)polyesters using lipases has been described in detail and a number of applications are currently being investigated [6, 7, 11–13]. Polyesters can be accessed via the three different routes depicted in Fig. 2. ROP of cyclic esters (route a, Fig. 2) have been thoroughly investigated by the groups of Kobayashi and Gross [8, 9, 11, 12]. In this case, no condensation product is formed, which allows, in principle, the preparation of high molecular weight polyesters. In polycondensation reactions of AA–BB-type monomers such as diacids or diesters and diols (route b, Fig. 2), or polycondensation reactions of AB-type monomers such as ester or acid alcohol monomers (route c, Fig. 2), a condensation product (water or an alcohol) is formed that needs to be removed in order to reach high conversions.

We will first focus on the implications of the catalytic activation mechanism of lipases, namely monomer activation, on the attainable chains lengths and polydispersity when cyclic esters are converted into polyesters. Moreover, the control of

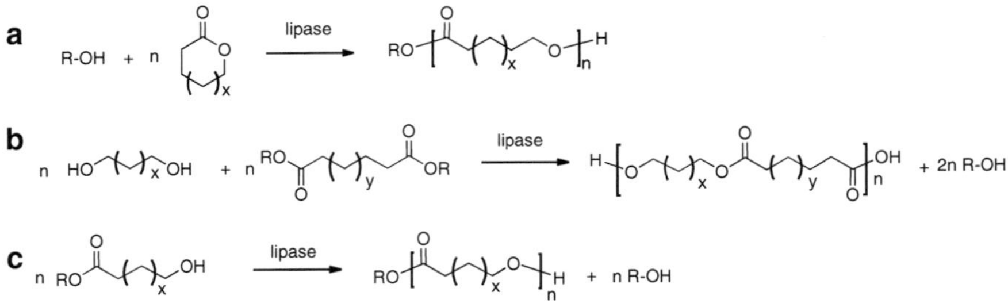

Fig. 2 Synthetic routes to procure polyesters using lipase catalysis: (**a**) ROP of cyclic esters, (**b**) polycondensation of AA–BB monomers, and (**c**) polycondensation of AB monomers

monomer sequence during ROP and the endgroup fidelity will be discussed. Then, we will briefly discuss the application of lipases in polycondensation reactions, concentrating on the production of multifunctional polyesters in a one-step approach.

3.1 Lipase-Catalyzed ROP of Cyclic Esters

3.1.1 Monomer Activation During ROP

In lipase-catalyzed ROP, it is generally accepted that the monomer activation proceeds via the formation of an acyl-enzyme intermediate by reaction of the Ser residue with the lactone, rendering the carbonyl more prone to nucleophilic attack (Fig. 3) [60–64, 94]. Initiation of the polymerization occurs by deacylation of the acyl-enzyme intermediate by an appropriate nucleophile such as water or an alcohol to produce the corresponding ω-hydroxycarboxylic acid or ester. Propagation, on the other hand, occurs by deacylation of the acyl-enzyme intermediate by the terminal hydroxyl group of the growing polymer chain to produce a polymer chain that is elongated by one monomer unit.

Mechanistic investigations of lipase-catalyzed ROP of unsubstituted lactones revealed that the polymerizations follow Michaelis–Menten kinetics and that the formation of the acyl-enzyme intermediate is the rate-determining step [62]. In substituted lactones, depending on the size of the substituent, the deacylation step can become rate-determining [65]. The determination of the Michaelis–Menten constant, K_M, and the maximal rate of polymerization, V_{max}, of unsubstituted lactones of varying ring size showed that K_M is relatively independent of the ring size, and that no obvious trend could be discerned for V_{max} [66, 67]. Interestingly, lactones possessing a *cisoid* ester bond conformation appeared less reactive in CALB-catalyzed reactions than did large ring lactones in a *transoid* conformation (Fig. 4a). This is an opposite trend to that of the chemical reactivity of lactones, where ring strain in the small *cisoid* lactone is the driving force for reaction and polymerization [96].

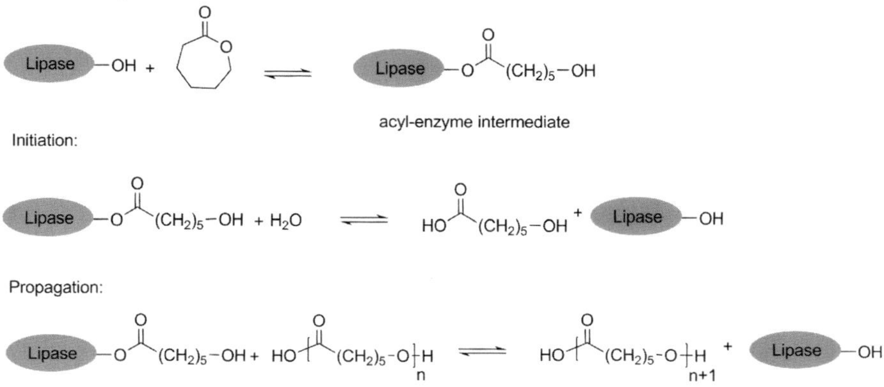

Fig. 3 Proposed mechanism for lipase-catalyzed ROP of lactones [63]

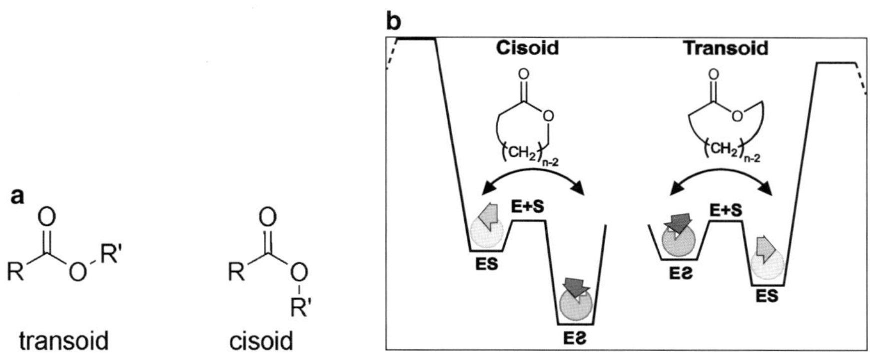

Fig. 4 (**a**) *Cisoid* and *transoid* conformation of the ester bond. (**b**) Productive and unproductive binding due to the *cisoid* or *transoid* nature of the ester bond in lactones in the CALB active site [95]

Research by Veld et al. to understand the reactivity of *cisoid* and *transoid* ester bonds in more detail focused on docking studies of small lactones (δ-valerolactone, VL, and ε-caprolactone, CL) and a medium-sized lactone (heptanolactone, HL) in the active site of CALB [95]. Although VL and CL possess a *cisoid* conformation of the ester bond in the ring, HL and all other large ring lactones and open chain esters (mostly) posses low-energy *transoid* ester bond conformations (Fig. 4a). For the *cisoid* lactones, competitive inhibition by wrongly bound substrate appeared to be responsible for the low experimentally observed reactivity (Fig. 4b). Hence, the reactivity of lactones in CALB-catalyzed reactions is governed by the *cisoid* or *transoid* nature of the ester bond. Despite the high chemical reactivity of *cisoid* ester bonds, the enzymatic reactivity of *cisoid* esters is much lower. This finding suggests that in a ROP of lactones to polyesters, in which a monomer-activated mechanism is operational, the polyester product can be more reactive in the activation step than the lactones substrate. This will have important implications for the ability to control the polydispersity, molecular weight, and endgroup fidelity in lipase-catalyzed ROP.

3.1.2 Control Over Polydispersity and Molecular Weight in ROP of Lactones

Numerous comparative studies on the ROP of CL in organic solvents have been conducted to enhance the mechanistic and kinetic understanding of lipase-catalyzed ROP. Investigated reaction parameters are, for example, the enzyme origin [62, 97–100], concentration of monomer [70], temperature [68, 69], organic solvent [61, 101], and water content [72, 73]. Enzymatic ROP reactions have also been investigated in alternative solvents such as supercritical media [102–104] and ionic liquids [105, 106].

CL is a particularly interesting monomer because it has a high ring strain, making it reactive in (chemical) ROP [107]. Its corresponding polymer, poly(ε-caprolactone) (pCL), has been evaluated in a number of biomedical applications. A key factor in the lipase-catalyzed ROP of CL is the ability to control the polydispersity and molecular weight during the polymerization reaction. Gross and coworkers proposed that the ROP of CL is pseudo-living, based on linearity of the ln(1-conversion) versus time plots and the linear increase of molecular weight with conversion [60]. However, the polydispersities (polydispersity index, PDI, was typically between 1.5 and 2.0) suggested a limited control during the polymerization. Moreover, at high substrate conversions (>80%), deviations from linearity appeared that were attributed to concomitant condensation reactions [64]. In fact, most papers dealing with lipase-catalyzed ROP report a PDI in the range of 2.0 for unprecipitated samples, showing that side reactions also occur during the ROP. The fact that a lipase can activate every ester bond, and preferably accepts *transoid* ester bonds as substrates, implies that transacylation reactions will always compete with monomer activation during the polymerization. In other words, competing transacylation reactions and chain transfer cannot be avoided during lipase-catalyzed polymerization reactions. A PDI of 2.0 is expected as a result of the statistically randomized chain length distribution.

Control over the molecular weight, on the other hand, has been achieved reasonably well. In the absence of water, high molecular weight polyesters are typically obtained [74, 75, 108], whereas increasing the water content (which acts as an initiator) leads to a higher number of polymer chains and thus lower molecular weight polyesters [64, 101]. Also, the use of initiators containing an alcohol allows the preparation polymers of predictable molecular weight [72, 76, 109, 110].

3.1.3 Endgroup Fidelity in Lipase-Catalyzed ROP of Lactones

A second key factor in the lipase-catalyzed ROP of CL is the ability to control the nature of the endgroups. This has important consequences in preparation of α,ω-functionalized polyesters, which are currently being explored in a variety of applications [111, 112]. A systematic study by Heise and coworkers on the initiating efficiency of different initiators (**1–3** in Fig. 5) in the CALB-catalyzed ROP of CL showed that three polymeric products were formed: cyclic species, water-initiated

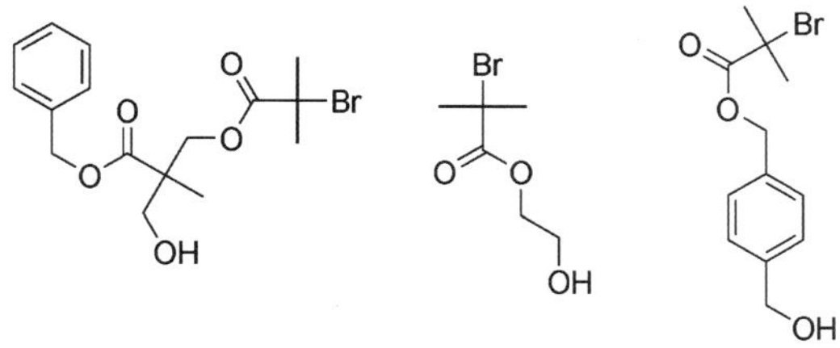

1 2 3

Fig. 5 Chemical structures of the initiators **1–3** investigated in the ROP of CL [113]

pCL chains, and pCL chains with the desired initiator attached [72]. Depending on the concentration of the monomer and the water activity of the reaction, the relative amount of the three products varied. The degree of pCL functionalization was at best 95%. A further reduction of water initiation was not feasible because some water is required for the activity of the enzyme [114]. A careful evaluation of the polymerization of CL in the presence of initiators **1–3** by a combination of liquid chromatography under critical conditions (LCCC), size exclusion chromatography (SEC) and Maldi-TOF-MS revealed that water, even when stringent drying conditions were applied, dominated the initial initiation process. However, depending on the exact reaction conditions, the initiator was incorporated into the polymer structure to a high degree as a result of transacylation reactions [113].

Interestingly, recent work by Hult and coworkers hypothesized a water tunnel in CALB that provides the active site with substrate water [115]. Any water present acts as a competing nucleophile when an alcohol is the desired nucleophile, and affords hydrolysis. To circumvent water as a favored nucleophile, polymerization reactions are usually performed under dry conditions in organic solvents or directly in bulk. However, absolutely dry conditions are difficult to maintain, thus it is of great interest to minimize CALB's preference for water as a nucleophile. A small, focused library created in order to prevent water from entering the active site through the tunnel was screened for increased transacylation over hydrolysis activity. One mutant, in which the inner part of the tunnel was blocked, catalyzed the transacylation of vinyl butyrate to butanol 14 times faster than hydrolysis. Such mutants with a blocked water tunnel could be very valuable in applications in which high molecular weight polyesters are required or excellent control over endgroup functionality is desired.

Careful selection of the reaction conditions, and working under thermodynamic control (i.e. perform reactions under equilibrium conditions), allows high endgroup fidelity in a lipase-catalyzed ROP, even in a one-pot procedure [58, 80, 111, 112, 116–118]. Initiation with mercapto-ethanol or 6-mercapto-1-hexanol in lipase-catalyzed ROP of CL and ω-pentadecanolactone (PDL), respectively, resulted in high initiator incorporation (95%). It was found that the alcohol acted as the initiating species, whereas the thiol was unreactive in the initiation step [80]. The latter is a result of the higher chemoselectivity of lipases for the alcohol group over the thiol group in the deacylation step. In situ end-capping of the polymers with thiobutyrolactone or 11-mercaptoundecanoic acid afforded dithiol end-capped polyesters in high yields and with a high degree of functionalization [24, 80, 117]. Vinyl(meth)acrylates were also found to be suitable end-cappers during the polymerization reactions [70, 119]. In combination with HEMA or HEA as an initiator, di(meth)acrylate macromonomers are accessible in a one-pot enzymatic procedure [58, 79, 81]. In addition, in a one-pot approach, glycidol initiation in combination with a divinyl ester initiator also resulted in epoxide end-capped polyesters that can be cured into biodegradable coatings [111].

3.1.4 Randomness in Copolymerizations of Lactones with Lipase-Catalyzed ROP

Because polyester formation during the ROP of cyclic lactones proceeds via a monomer activation mechanism, polymerizing a mixture of lactones should result in the formation of random copolyesters. Indeed, as observed by several groups, copolymerization of CL and PDL [120], CL and VL [121], PDL and 2-oxo-12-crown-4 (OC) [122, 123], CL and OC [124], PDL and *p*-dioxanone (DO) [125], and CL and 1,5-dioxepan-2-one (DXO) [126] afforded random copolymers, as evidenced by differential scanning calorimetry, ^{1}H-NMR, ^{13}C-NMR, and solid state NMR. For example, in the copolymerization of PDL and OC, the rate of PDL polymerization was five times lower than the rate of OC polymerization [123]. Characterization of the obtained copolymers with ^{1}H-NMR showed a random structure of the copolymer, as inferred from the theoretical and experimental dyad fractions. This has enormous potential for application of enzymatically prepared copolymers since the melting point, crystallinity, polarity, and degradability of the polyesters can be carefully tuned by simple mixing of two monomers [124, 127]. On the other hand, when polyesters with a block or gradient copolymer structure are desired, other (chemical) catalysts are more suitable, or a combination of different polymerization techniques can be considered [10].

3.1.5 Polymerizability of Cyclic Diesters with Lipases

In addition to lactones, lactides and glycolides are highly interesting monomers for evaluation in lipase-catalyzed ROP. Lactide is the cyclic dimer of lactic acid and can occur in three stereo-configurations: L-lactide, D–lactide, and D,L- or *meso*-lactide (Fig. 6). In analogy, glycolide is the cyclic dimer of glycolic acid but is

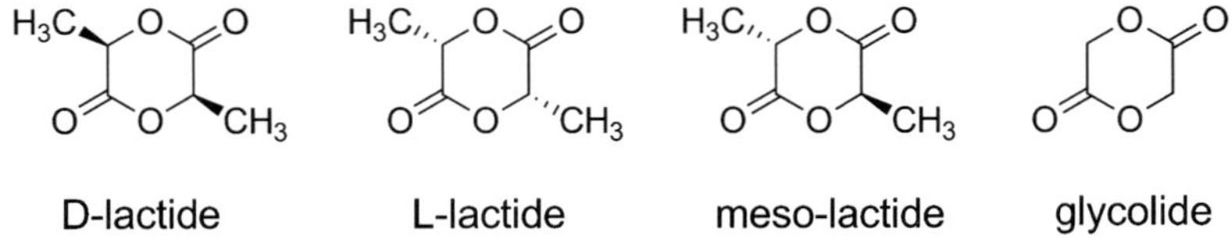

Fig. 6 Chemical structure of different lactide stereomers and glycolide

achiral because it lacks a stereogenic centre (Fig. 6). Both lactide and glycolide are highly reactive in chemical ROP and afford biocompatible and biodegradable polymers that are used in a variety of biomedical applications. This makes both monomers interesting for exploration in lipase-catalyzed ROP, especially when the lipase is CALB since this has a high polymerization activity. However, there are two problems that could limit the application of CALB in the ROP of lactides. First, CALB is known to be a slow catalyst when ester substrates have a substituent at the α-position (next to the carbonyl) [128] and, secondly, CALB is highly enantioselective for *R*-secondary alcohols in the deacylation step [27, 129]. This would imply that polymerizations of L-lactide, which give an *S*-secondary alcohol after ring-opening, will be problematic.

The reactivity of lactic acid as an acyl donor and an alcohol nucleophile was investigated in detail by Adlercreutz and coworkers [130]. The esterification of lactic acid with a variety of alcohols in hexane was very promising, employing Novozym 435 as the biocatalyst. Moreover, it was found that in the esterification of (*rac*)-lactic acid, both enantiomers showed similar reactivity, indicating that CALB is not enantioselective in the acylation step when the substituent is placed at the carbonyl side. Hence, lactic acid is readily activated by Novozym 435. Interestingly, no lactic acid oligomers were formed, even when an excess of lactic acid was used, which suggests that the secondary alcohol is rather unreactive in the deacylation step. This was corroborated when fatty acids were employed as acyl donor in an esterification of the alcohol in lactic acid: no ester formation was observed. Apparently, the carbonyl group adds additional sterical hindrance and/or changes the nucleophilicity of the alcohol, which dramatically decreases the reactivity of the secondary alcohol in lactic acid. However, on longer time scales, esterification of lactic acid with caprylic acid in hexane was observed when employing Novozym 435, but the yields were low and the reaction times long (30% conversion in 4 days) [131].

The lack of reactivity of the alcohol moiety of lactic acid when using Novozym 435 as catalyst implies that lactide is not a good substrate in enzymatic ROP because propagation requires a reactive chain end that deacylates the enzyme-activated monomer. Indeed, copolymerizations of lactide with CL showed a dramatic decrease of CL consumption after an initial rapid consumption of lactide, which appeared to be the better substrate. Many weeks were required in order to observe the formation of random copolymers [132]. In contrast, D-lactide was polymerized within 2 days using Novozym 435, but a proper selection of reaction conditions was crucial in order to obtain good molecular weights [133]. A recent report also claims the successful polymerization of (*rac*)-lactic acid in a polycondensation reaction, although the yields and molecular weights were very low [134].

Use of *Pseudomonas cepacia* lipase (lipase PS) or *Porcine* pancreatic lipase does allow for the enzymatic ROP of lactide. Matsumura and coworkers reported polymers with extraordinarily high molecular weights (M_w up to 270 kDa) and very narrow PDI (<1.3) [135–137]. However, high temperatures (130 °C) were needed to achieve good conversions, and polymerizations proceeded only when conducted in bulk. It is conceivable that another non-enzymatic mechanism contributed in these polymerizations. In fact, Koning and coworkers synthesized copolymers of glycolide and lactide using immobilized *P. cepacia* lipase and pointed out the possibility of a cationic mechanism being operational during the enzymatic polymerization [138].

Despite the fact that glycolic acid has been successfully used as an acyl donor in esterification reactions with fatty alcohols, there are few reports dealing with the enzymatic ROP of glycolide [139]. On the other hand, cyclic diesters based on ethylene glycol have been polymerized successfully by lipase catalysis and afford AA–BB-type polyesters [140, 141].

3.2 Lipase-Catalyzed Polycondensation Reactions

Polycondensation reactions of AA–BB- and AB-type monomers (Fig. 2) using lipases have been described in great detail by a number of groups [32–39, 41, 44, 45, 142–146], and pioneering work by Baxenden Chemicals showed the feasibility of performing this process on a large scale [142]. Polycondensations are equilibrium reactions requiring removal of the formed condensation product to shift the equilibrium to high conversion. The equilibrium conversion of an ester from an alcohol and an acid is typically 70%, thus, the formed water needs to be removed effectively to drive the reaction to completion. Since lipases are catalysts, the position of the equilibrium is not altered, although it has been observed that the equilibrium conversion of immobilized preparations can be higher than 70% as a result of the altered water activity when resins are present [147, 148].

Methods to achieve high conversions in (trans)esterifications involve the use of activated esters such as vinylesters [46, 49] (the formed vinyl alcohol tautomerises into an aldehyde or ketone) or the use of vacuum [36, 37] or molsieves [47] to remove the condensation product and thus shift the equilibrium to high conversion. The most frequently applied lipase in polycondensation reactions is CALB. CALB is well-known for its deep and narrow active site and high degree of (enantio)selectivity (see [10] for details on the enantioselectivity of CALB) [16, 17]. Although the substrate scope of Novozym 435 is quite broad, phenols [149] or ester functionalities with a substituent at the α-position [128] are poorly accepted. On the other hand, benzylic [150] or aliphatic alcohols, aliphatic esters and acids, and even aromatic esters [48] are excellent substrates that have led to a large variety of (co)polyesters.

Important advantages of lipases compared to catalysts such as Lewis or Brønsted acids are the low reaction temperatures, which permit thermally labile monomers to be applied, and their high regioselectivity, which allows multifunctional monomers to be directly polymerized in high selectivity and thus avoids the necessity of protective group chemistry. The use of multifunctional substrates (Fig. 7) in particular has received a lot of attention because the regioselective nature of lipases allows the preparation of hydroxy- [40, 50, 51, 143, 145, 151–156] or carboxylic-acid-functional [146] polyesters that have interesting applications as biomedical and biodegradable materials [152]. For example, the copolymerization of sorbitol and

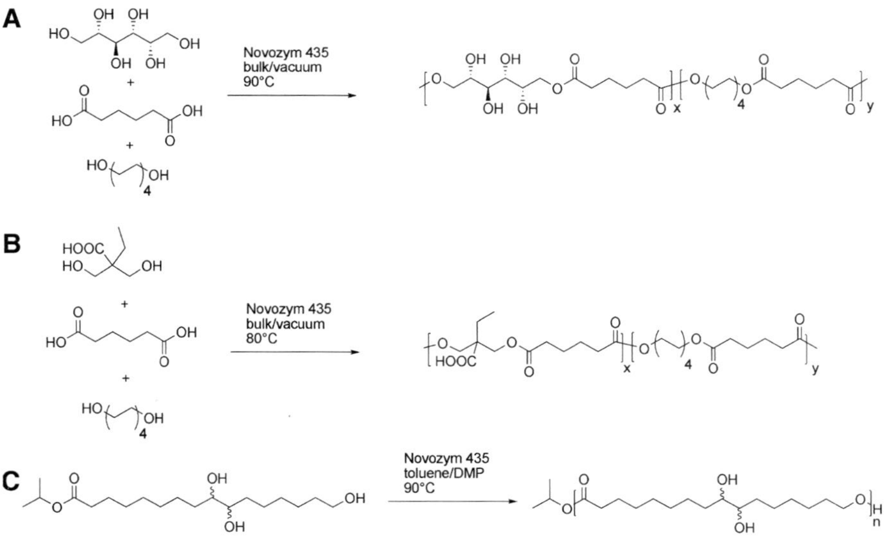

Fig. 7 Different approaches (*A-C*) to the direct formation of hydroxy- or carboxylic-acid-functional polyesters using Novozym 435

adipic acid using Novozym 435 in bulk, and using vacuum to remove the formed water, afforded water-soluble polymers with an M_n of 10.6 kDa and a PDI of 1.6. The reaction occurred predominantly at the primary alcohol groups, with a regioselectivity of 85% [40]. Adding 1,8-octanediol as a comonomer increased the M_n to 34 kDa and the regioselectivity to 95% (reaction A in Fig. 7). Thermal analysis showed that these copolymers were semicrystalline, low melting, soft materials [40].

Terpolymerizations of bis(hydroxymethyl)butyric acid (BHB), 1,8-octanediol, and adipic acid in bulk showed that Novozym 435 was strictly selective for esterification of BHB hydroxyl groups while leaving the carboxylic acid unchanged [146]. Copolyesters with 9–45 mol% BHB-adipate units were formed with M_w values ranging from 21 to 2.3 kDa. This direct polymerization without protection–deprotection chemistry afforded linear aliphatic copolyesters with pendant free acid groups (reaction B in Fig. 7). Interestingly, hyperbranched aliphatic copolyesters were prepared by Frey and coworkers, and copolymerization of BHB and CL using Novozym 435 in toluene at 85 °C showed that the acidic group in BHB can be reactive under the proper conditions [157]. This combination of ROP and polycondensation allowed control of the degree of branching between 0 and 0.3, and polymers were obtained with M_n of 61.5 kDa and PDI of 1.2. The degree of branching permits the fine-tuning not only of the crystallinity and melting points but also of rheological properties, which is interesting for future applications.

The polymerization of isopropyl aleuritate with Novozym 435 afforded linear polymers of good molecular weight (M_n 5.6 kDa, PDI 3.2) [51]. In this case, the regioselectivity was close to 100%, a result of the sterically hindered secondary alcohols (reaction C in Fig. 7). The thermal properties of the polymers could be tuned by copolymerization with CL and the degree of branching remained close to zero, showing that these secondary alcohols are unreactive. Subsequent reaction of the secondary alcohols with hexylisocyanate showed that these polyesters can be readily post-modified by suitable reagents.

4 Polycarbonates

Cyclic carbonates are an interesting class of monomers because polycarbonates such as poly(trimethylene carbonate) are often amorphous and show well-defined degradation properties, making them suitable for a variety of biomedical applications. ROP of cyclic carbonates is typically conducted with metal-based catalysts such as $Sn(Oct)_2$ but lipases were also found to be active in the enzymatic ROP of a variety of substituted and unsubstituted cyclic carbonates (Fig. 7). Trimethylene carbonate (TMC) was the first monomer to be evaluated in enzymatic ROP [83–85]. Novozym 435 gave a rapid polymerization reaction, and an M_n of 15 kDa was reached with complete monomer conversion within 120 h at 70 °C [85]. Higher molecular weights have been reached by using lipase PPL [83] (M_n 169 kDa, PDI 3.5) and PPL immobilized on silica nanoparticles (M_n 48 kDa, PDI 2.0) [86] although this required a temperature of 100 °C. Other cyclic carbonates such as hexamethylene carbonate (HMC) [87] or substituted carbonates 1-MeTMC [90] and 2,2-diMeTMC [158] (Fig. 8) have also been successfully polymerized into polymers of good molecular weight. The use of protected functional groups at the TMC ring, such as the 5-methyl-5-benzyloxycarbonyl-1,3-dioxan-2-one (MBC) [82, 88], 5-benzyloxy-trimethylene carbonate (BTMC) [89], or tartaric-acid-derived cyclic carbonates (TAC) (Fig. 8), allows for easy access to the hydroxy-functional or carboxy-functional polycarbonates, which may be used to tune the degradation behavior of these polymers.

Lipases not only catalyze ROP but also catalyze transacylation reactions. This was exploited by preparing random copolymers of TMC with CL or PDL [91, 92, 159]. The PDL–TMC poly(carbonate-esters) show co-crystallization behavior of

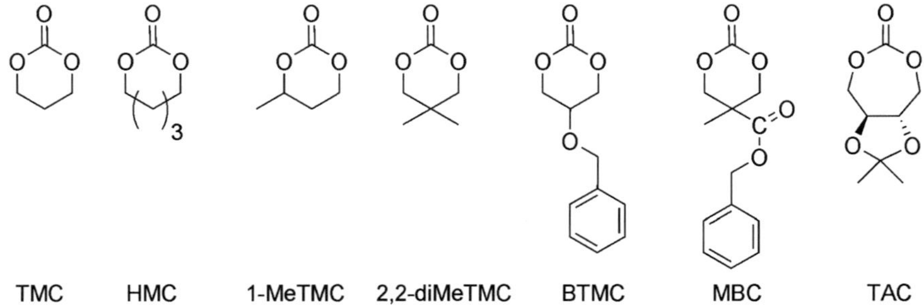

Fig. 8 Chemical structures of cyclic carbonates explored in lipase-catalyzed ROP

their monomers in a unit cell, which facilitates tuning the degradation behavior without compromising on the thermal and mechanical properties of the copolymers [159].

Cyclic carbonates are not commercially available and have to be synthesized prior to use. As a result, commercially available carbonates such as diethyl carbonate [55–57] or diphenyl carbonate [93] were evaluated in polycondensation reactions with diols to prepare polycarbonates since they allow a broader spectrum of polymers to be accessed. Unfortunately, polymerizations employing diethyl carbonate require the use of an excess diethyl carbonate [55]. Nevertheless, polymers with molecular weight of 40 kDa were achieved within 16 h. Also, the polymerization of diphenyl carbonate with butane-1,4-diol or hexane-1,6-diol via the formation of a cyclic dimer produced polymers with molecular weights ranging from 119 to 339 kDa [93].

5 Polyamides

Polyamides are a very important class of polymers with everyday applications in textiles, automotive materials, and plastics for electronics. Typically, polyamides are chemically prepared by ROP of lactams or the polycondensation of diamines and diacids. Although ester and carbonate bonds are excellent substrates for lipases, amide bonds are not. The hydrolysis of an amide bond is usually impossible using lipases because of the higher activation energy required for the formation of the enzyme-bound tetrahedral transition state of amides in comparison to the more polar ester substrates [160]. This is attributed to the electron delocalization present in the amide bond, which increases resonance stabilization in the ground state. Nevertheless, structural features that prevent resonance stabilization can activate amide substrates towards electrophilic attack by water. Thus, strained amides present in β-lactam (four-membered) rings are readily hydrolyzed using lipases [23, 161, 162]. Alternatively, the reaction of an ester and an amine (a reaction that is exploited industrially in the kinetic resolution of racemic amines) also gives amide products [163–165]. Interestingly, amines are much poorer nucleophiles than alcohols in a lipase-catalyzed reaction, which significantly reduces the reaction rate. For example, in a kinetic resolution of 1-phenylethanol with methyl butyrate the rate is approximately 130 times higher than in a kinetic resolution of 1-phenylethanamine (Table 2). Introduction of an oxygen at the β-position of the ester bond resulted in a 100-fold rate enhancement for the kinetic resolution of the amine but not for the alcohol. A structural basis for this effect was recently given by Hult and coworkers [28]. A molecular modeling study of the transition-state analog for the aminolysis showed that an interaction between the β-oxygen atom in methoxyacetate and the amine nitrogen atom could be a key factor in the rate enhancement. This interaction presumably stabilizes the transition state and increases the reaction rate.

Despite the feasibility of making and breaking amide bonds using lipases in small molecules, the use of lipases in polyamide synthesis is still very limited. The prime

Table 2 Initial reaction rates and normalized reaction rates (between brackets) of alcohols and amines using a normal acyl donor or a β-oxygen-activated acyl donor [28]

Acyl Donor \ Substrate	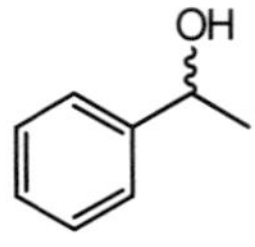	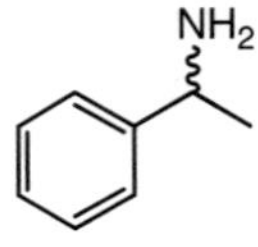
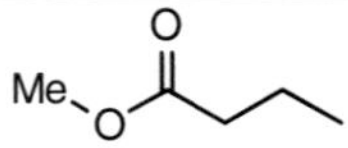	7.8 (1)	0.06 (1)
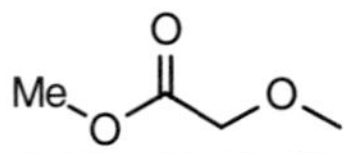	8.1 (1)	13 (200)

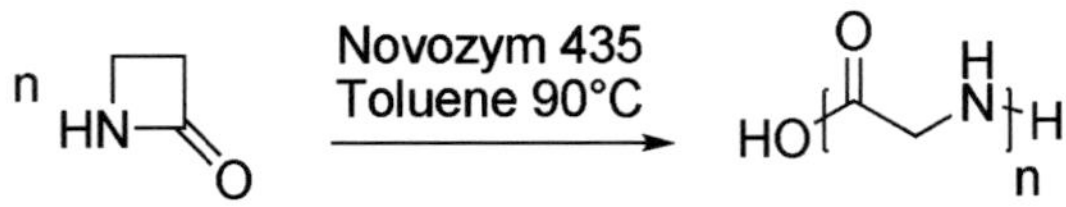

Fig. 9 Polymerization of β-lactam into poly(β-alanine) with Novozym 435

reason is presumably the poor solubility of polyamides in the (apolar) solvents that are typically applied in lipase-catalyzed reactions. Recently, some very elegant approaches have appeared in the literature and show that the preparation of well-defined polyamides by lipase catalysis is feasible.

The synthesis of poly(β-alanine) from unsubstituted β-lactam was achieved by Loos and coworkers using Novozym 435 as the catalyst at 90 °C in anhydrous toluene [22]. The polymer formation proceeded by ring opening of the β-lactam, after which oligomers were built up by reaction of the amine chain end with an activated monomer (Fig. 9). A control reaction with deactivated enzyme resulted in less than 5% product, showing the necessity of the lipase in the monomer activation step. Maldi-TOF-MS showed a distribution with peaks up to 18 monomeric units, and both cyclic and linear species were found. From ^{1}H-NMR, an average degree of polymerization of eight could be derived, which was limited by the solubility of the formed polymer in the reaction medium.

In order to keep polyamides soluble in relatively apolar solvents, the use of flexible (macro)monomers such as α, ω-(diaminopropyl)polydimethylsiloxane [52] or oligoethyleneglycol-based diamines [53, 54] has been proven to be a successful approach (Fig. 10). Polycondensations of dimethyl adipate with a variety of diamines were successful in bulk and at moderate temperatures between 60 and 100 °C (reaction A in Fig. 10). The low temperatures (60–100 °C) that suffice in these polymerizations also allow the use of monomers that are thermally instable, such as diethyl fumarate [53]. Moreover, multifunctional amines could be regioselectively polymerized up to molecular weights of 9 kDa, making lipase catalysts a valuable tool for the preparation of well-defined polyamides that can be further functionalized with active groups.

For the silicon-containing polyamides, molecular weights up to 9.4 kDa were achieved (reaction B in Fig. 10). Copolymerizations of 1,8-octanediol,

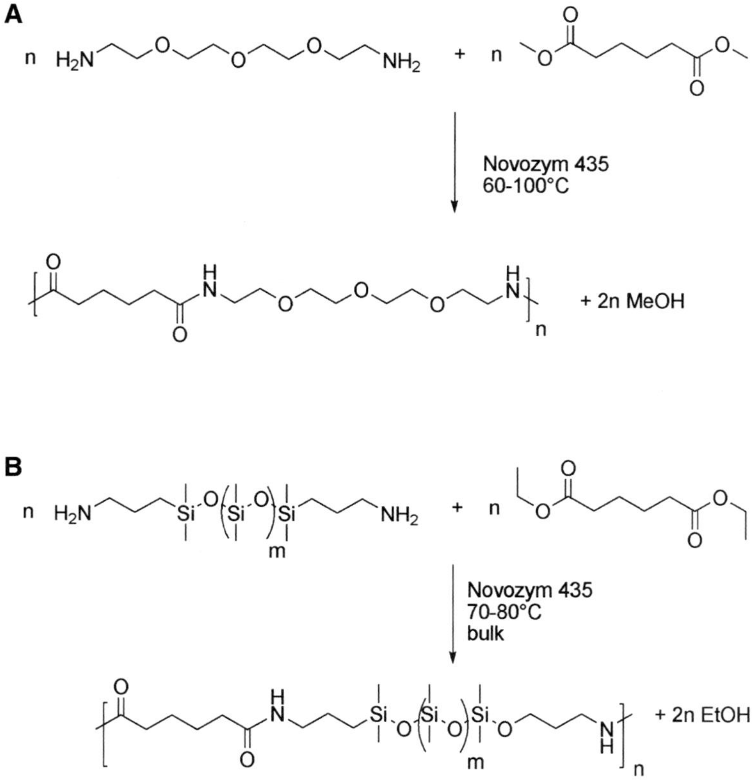

Fig. 10 Formation of polyamides (*A*) and silicon-containing polyamides (*B*) using Novozym 435 catalysis

α,ω-(diaminopropyl)polydimethylsiloxane and diethyl adipate using Novozym 435 at 70 °C in bulk showed that the formation of amide bonds occurred more rapidly than the formation of ester bonds [52]. This resulted in copolymers with a tendency to form a block-like sequence distribution. The relative amount of amide units compared to ester units in the copolymers has a strong effect on the physical properties. A high content of ester bonds resulted in hard, solid materials whereas a high content of amides, and thus silicon, produced sticky materials.

6 Polythioesters

Aliphatic polyesters have received great interest for potential biomedical application. Polythioesters (polyesters in which one of the oxygen atoms of the ester groups has been replaced by a sulfur atom) have received less attention, although these materials are expected to show interesting material properties such as higher melting points and refractive indices than the corresponding polyesters [166, 167]. Moreover, some microorganisms are known to form (co)polymers based on β-mercapto propionic acid and β-hydroxy propionic acid, which has renewed interest in this class of polymers [168–171].

Some examples of lipase-catalyzed polythioester synthesis have been mentioned in the literature, although thioesters and thiol groups generally possess a lower reactivity than the corresponding carboxylic esters and hydroxy nucleophiles in lipase-catalyzed reactions. The lower reactivity has been demonstrated by the O-selective acylation of ω-mercapto alkanols [172] and the possibility to selectively end-cap polymers with thiol groups with a high degree of efficiency, as discussed before [80]. The lower reactivity of thiol nucleophiles in CALB-catalyzed reactions necessitates the use of relatively large amounts of enzyme, high reaction temperatures, and long reaction times for the preparation of polythioesters. However, some examples are known in which lipases showed a high reactivity towards thiol nucleophiles [31], thioesters [173], and thioacids [174].

CALB-catalyzed copolymerization of CL with 11-mercaptoundecanoic acid (11MU) leads to the formation of poly(ester-*co*-thioester)s having a M_n of 13.7 kDa (PDI 1.6) after precipitation [30] (Fig. 11). The amount of incorporated 11MU (8.7 mol%) was slightly less than the feed ratio (10 mol%). Similar results were obtained when using 3-mercaptopropionic acid (3MP) as a comonomer (M_n 14.3 kDa, PDI 1.4). CALB-catalyzed transesterification of pCL with either 11MU or 3MP resulted in similar ^{1}H-NMR and ^{13}C-NMR spectra as the direct copolymerization of the two monomers, showing that continuous transesterification plays an important role in the microstructure of the polymer [30].

Later, polythioesters were prepared by the polycondensation of dicarboxylic acid diesters with a dithiol [43] (Fig. 12). Initially, various lipases were screened in the polycondensation between diethyl adipate and hexane-1,6-dithiol at 120°C in the presence of molecular sieves. The use of 70 wt% immobilized CALB as catalyst proved to be efficient in these polycondensation reactions with dicarboxylic acid

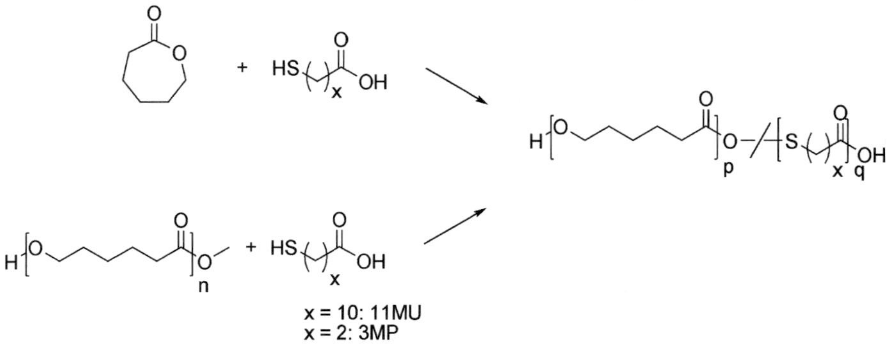

Fig. 11 Copolymerization of CL with aliphatic ω-mercapto carboxylic acids (*top*) and transesterification of pCL with aliphatic ω-mercapto carboxylic acids (*bottom*) to give random copolymers with thioester linkages [30]

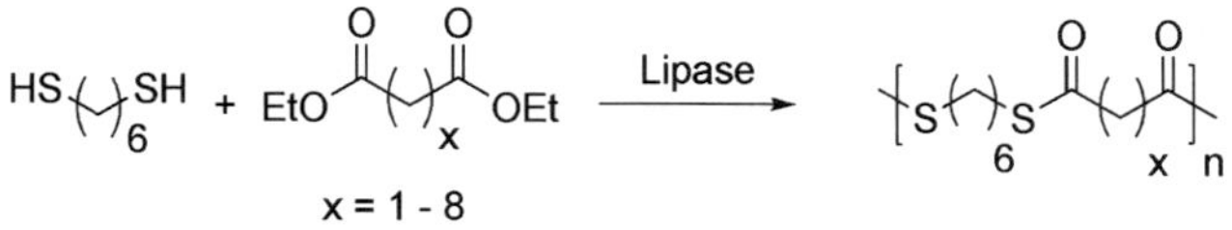

Fig. 12 Lipase-catalyzed polycondensation of 1,6-hexanedithiol with dicarboxylic acid diesters to give polythioesters [43]

diesters of varying spacer length with hexane-1,6-dithiol. Low molecular weight polythioesters were obtained in 75–90% yield (M_n 3.7–6.0 kDa, PDI 1.7–2.0). A higher molecular weight polythioester (M_n 50 kDa, PDI 2.3) was obtained by preparation of cylic, oligomeric polythioesters under dilute conditions, followed by CALB-catalyzed ROP in bulk at 120°C for 2 days [25, 43].

Imbalance in the stoichiometry of polycondensation reactions of AA–BB-type monomers can be overcome by changing to heterofunctional AB-type monomers. Indeed, 11MU has been subjected to bulk polycondensation using lipases as catalyst in the presence of 4 Å molecular sieves. At 70°C, CALB showed 84% monomer conversion and a low molecular weight polymer (M_n 1.1 kDa, PDI 1.9). No significant polymerization was observed with other lipases (except *P. cepacia* lipase, 47% conversion, oligomers only) and in reference reactions with thermally deactivated CALB or in the absence of enzyme. Further optimization of the reaction conditions (60 wt% CALB, 110°C, 3 days, 4 Å molecular sieves) gave a polymer with M_n of 14.8 kDa (PDI 2.3) in 86% yield after precipitation [42].

7 Conclusions

Lipase-catalyzed polymerizations have found a place in the field of synthetic polymer chemistry. Detailed research conducted in the past decade has brought the understanding of this novel technology to a level that is known in traditional polymerization techniques. This led to a situation where lipase-catalyzed ROP and polycondensation can be compared directly to its chemical counterpart, and its specific advantages identified. The recent breakthroughs in enzymatic ROP are to a large extent due to Novozym 435. The commercial availability of this stable, robust immobilized lipase formulation has been very important in bridging the worlds of biocatalysis and polymer chemistry. It has also allowed polymer chemists to routinely apply biocatalysis without the necessity of a deep knowledge of enzymology or biotechnology.

Nevertheless, the limitations of Novozym 435 have also been clearly identified. Water is the preferred nucleophile, making 100% end-functionalization very challenging, while the preference of CALB for *transoid* ester bonds limits the potential to reach low polydispersities in a ROP. Multidisciplinary projects focusing both on the enzymology side of the biocatalyst (e.g., by improving the lipase by mutations) and on the polymers required for specific applications can in the near future lead to a paradigm shift in the field of enzymatic polymerizations. Similar to the great advances made in bioorganic chemistry (in which biocatalytic routes are now routinely applied in chemical conversions to prepare optically active intermediates at lower cost and with less waste and less energy), a close collaboration between the biological and the materials world will allow the efficient development of smart materials and polymer architectures with a high level of structural complexity and functions.

References

1. Dechy-Cabaret O, Martin-Vaca B, Bourissou D (2004) Chem Rev 104:6147
2. Dove AP (2008) Chem Commun:6446
3. Jérôme C, Lecomte P (2008) Adv Drug Deliv Rev 60:1056
4. Mecerreyes D, Jérôme R, Dubois P (1999) Adv Polymer Sci 147:1
5. Kamber NE, Jeong W, Waymouth RM, Pratt RC, Lohmeijer BGG, Hedrick JL (2007) Chem Rev 107:5813
6. Albertsson A-C, Varma IK (2002) Adv Polym Sci 157:1
7. Albertsson A-C, Varma IK (2003) Biomacromolecules 4:1466
8. Gross RA, Kumar A, Kalra B (2001) Chem Rev 101:2097
9. Kobayashi S, Uyama H, Kimura S (2001) Chem Rev 101:3793
10. Heise A, Palmans ARA (2010) Adv Polym Sci. doi: 10.1007/12_2010_74
11. Kobayashi S (2009) Macrom Rapid Commun 30:237
12. Kobayashi S, Makino A (2009) Chem Rev 109:5288
13. Albertsson AC, Srivastava RK (2008) Adv Drug Deliv Rev 60:1077
14. Nardini M, Dijkstra BW (1999) Curr Opin Struct Biol 9:732
15. Anderson EM, Larsson KM, Kirk O (1998) Biocatal Biotransform 16:181
16. Uppenberg J, Hansen MT, Patkar S, Jones TA (1994) Structure 2:293
17. Uppenberg J, Oehrner N, Norin M, Hult K, Kleywegt GJ, Patkar S, Waagen V, Anthonsen T, Jones TA (1995) Biochemistry 34:16838
18. Uppenberg J, Patkar S, Bergfors T, Jones TA (1994) J Mol Biol 235:790
19. Zaks A, Klibanov AM (1984) Science 224:1249
20. Pozo M, Gotor V (1993) Tetrahedron 49:10725
21. Pozo M, Gotor V (1995) Tetrahedron Asym 6:2797
22. Schwab LW, Kroon R, Schouten AJ, Loos K (2008) Macromol Rapid Commun 29:794
23. Park S, Forro E, Grewal H, Fueloep F, Kazlauskas RJ (2003) Adv Synth Catal 345:986
24. Hedfors C, Ostmark E, Malmström E, Hult K, Martinelle M (2005) Macromolecules 38:647
25. Kato M, Toshima K, Matsumura S (2007) Biomacromolecules 8:3590
26. Wen J, Zhuo RX (1998) Macromol Rapid Commun 19:641
27. Rotticci D, Haeffner F, Orrenius C, Norin T, Hult K (1998) J Mol Catal B Enzym 5:267
28. Cammenberg M, Hult K, Park S (2006) ChemBioChem 7:1745
29. van Rantwijk F, Sheldon RA (2004) Tetrahedron 60:501
30. Iwata S, Toshima K, Matsumura S (2003) Macromol Rapid Commun 24:467
31. Zaks A, Klibanov AM (1985) Proc Natl Acad Sci USA 82:3192
32. Mahapatro A, Kalra B, Kumar A, Gross RA (2003) Biomacromolecules 4:544
33. Mahapatro A, Kumar A, Gross RA (2004) Biomacromolecules 5:62
34. Mahapatro A, Kumar A, Kalra B, Gross RA (2004) Macromolecules 37:35
35. Sahoo B, Bhattacharya A, Fu H, Gao W, Gross RA (2006) Biomacromolecules 7:1042
36. Binns F, Harffey P, Roberts SM, Taylor A (1999) J Chem Soc Perkin Trans 1:2671
37. Binns F, Roberts SM, Taylor A, Williams CF (1993) J Chem Soc Perkin Trans 1:899
38. Duwensee J, Wenda S, Ruth W, Kragl U (2010) Org Proc Res Dev 14:48

39. Azim H, Dekhterman A, Jiang Z, Gross RA (2006) Biomacromolecules 7:3093
40. Kumar A, Kulshrestha AS, Gao W, Gross RA (2003) Macromolecules 36:8219
41. Dong H, Wang H-D, Cao S-G, Shen J-C (1998) Biotechnol Lett 20:905
42. Kato M, Toshima K, Matsumura S (2005) Biomacromolecules 6:2275
43. Kato M, Toshima K, Matsumura S (2006) Macromol Rapid Commun 27:605
44. Wallace JS, Morrow CJ (1989) J Pol Sci A Polym Chem 27:3271
45. Morrow CJ, Wallace JS (1990) In: Abramowicz DA (ed) Biocatalysis. Van Nostrand Reinhold, New York, p 25
46. Namekawa S, Uyama H, Kobayashi S (2000) Biomacromolecules 1:335
47. Mezoul G, Lalot T, Brigodiot M, Marechal E (1995) J Pol Sci A Polym Chem 33:2691
48. Mezoul G, Lalot T, Brigodiot M, Marechal E (1996) Polym Bull 36:541
49. Uyama H, Yaguchi S, Kobayashi S (1999) Polym J 31:380–383
50. Uyama H, Inada K, Kobayashi S (1999) Macromol Rapid Commun 20:171
51. Veld MAJ, Palmans ARA, Meijer EW (2007) J Pol Sci A Polym Chem 45:5968
52. Sharma B, Azim A, Azim H, Gross RA (2007) Macromolecules 40:7919
53. Gu Q-M, Maslanka MM, Cheng HN (2008) ACS Symp Ser Polym Biocatal Biomater 999:309
54. Gu Q-M, Maslanka WW, Cheng HN (2006) Polym Prepr 47:234
55. Matsumura S, Harai S, Toshima K (2000) Macromol Chem Phys 201:1632
56. Jiang Z, Liu C, Gross RA (2008) Macromolecules 41:4671
57. Matsumura S, Harai S, Toshima K (1999) Proc Jpn Acad B Phys 75:117
58. Xiao Y, Takwa M, Hult K, Koning CE, Heise A, Martinelle M (2009) Macromol Biosci 9:713
59. Bisht KS, Henderson LA, Gross RA, Kaplan DL, Swift G (1997) Macromolecules 30:2705
60. Henderson LA, Svirkin YY, Gross RA, Kaplan DL, Swift G (1996) Macromolecules 29:7759
61. MacDonald RT, Pulapura SK, Svirkin YY, Gross RA, Kaplan DL, Akkara J, Swift G, Wolk S (1995) Macromolecules 28:73
62. Namekawa S, Suda S, Uyama H, Kobayashi S (1999) Int J Biol Macromol 25:145
63. Uyama H, Namekawa S, Kobayash S (1997) Polym J 29:299
64. Mei Y, Kumar A, Gross R (2003) Macromolecules 36:5530
65. Peeters JW, van Leeuwen O, Palmans ARA, Meijer EW (2005) Macromolecules 38:5587
66. Duda A, Kowalski A, Penczek S, Uyama H, Kobayashi S (2002) Macromolecules 35:4266
67. van der Mee L, Helmich F, De Bruijn R, Vekemans JAJM, Palmans ARA, Meijer EW (2006) Macromolecules 39:5021
68. Kumar A, Gross RA (2000) Biomacromolecules 1:133
69. Mei Y, Kumar A, Gross RA (2002) Macromolecules 35:5444
70. Cordova A, Iversen T, Hult K (1999) Polymer 40:6709
71. Cordova A, Hult A, Hult K, Ihre H, Iversen T, Malmström E (1998) J Am Chem Soc 120:13521
72. de Geus M, Peeters J, Wolffs M, Hermans T, Palmans ARA, Koning CE, Heise A (2005) Macromolecules 38:4220
73. Matsumoto M, Odachi D, Kondo K (1999) Biochem Eng J 4:73
74. van der Meulen I, de Geus M, Antheunis H, Deumens R, Joosten EAJ, Koning CE, Heise A (2008) Biomacromolecules 9:3404
75. de Geus M, van der Meulen I, Goderis B, van Hecke K, Dorschu M, van der Werff M, Koning CEH, Heise A (2010) Polym Chem. doi:10.1039/b9py00360f
76. Panova AA, Kaplan DL (2003) Bioeng Biotechnol 84:103
77. de Geus M, Peters R, Koning CE, Heise A (2008) Biomacromolecules 9:752
78. Takwa M, Xiao Y, Simpson N, Malmström E, Hult K, Koning CE, Heise A, Martinelle M (2008) Biomacromolecules 9:704
79. Takwa M, Hult K, Martinelle M (2008) Macromolecules 41:5230
80. Takwa M, Simpson N, Malmström E, Hult K, Martinelle M (2006) Macromol Rapid Commun 27:1932
81. Uyama H, Suda S, Kobayashi S (1998) Acta Polym 49:700
82. Al-Azemi TF, Harmon JP, Bisht KS (2000) Biomacromolecules 1:493
83. Matsumura S, Tsukada K, Toshima K (1997) Macromolecules 30:3122
84. Kobayashi S, Kikuchi H, Uyama H (1997) Macromol Rapid Commun 18:575

85. Bisht KS, Svirkin YY, Henderson LA, Gross RA, Kaplan DL, Swift G (1997) Macromolecules 30:7735
86. Feng J, He F, Zhuo R (2002) Macromolecules 35:7175
87. Jiang ZZ, Liu C, Xie WC, Gross RA (2007) Macromolecules 40:7934
88. Al-Azemi TF, Bisht KS (1999) Macromolecules 32:6536
89. Feng J, Zhuo R, He F, Wang X (2003) Macromolecular Symp 195:237
90. Tasaki H, Toshima K, Matsumura S (2003) Macromol Biosci 3:4361
91. Kumar A, Garg K, Gross RA (2001) Macromolecules 34:3527
92. Deng F, Gross RA (1999) Int J Biol Macromol 25:153
93. Yamamoto Y, Kaihara S, Toshima K, Matsumura S (2009) Macromol Biosci 9:968
94. Cordova A, Iversen T, Hult K, Martinelle M (1998) Polymer 39:6519
95. Veld MAJ, Fransson L, Palmans ARA, Meijer EW, Hult K (2009) ChemBioChem 10:1330
96. Huisgen R, Ott H (1959) Tetrahedron 6:253
97. Kobayashi S, Takeya K, Suda S, Uyama H (1998) Macromol Chem Phys 199:1729
98. Noda S, Kamiya N, Goto M, Nakashio F (1997) Biotechnol Lett 19:307
99. Nakaoki T, Mei Y, Miller L, Kumar A, Kalra B, Kirk O, Christensen M, Gross RA (2005) Ind Biotechnol 1:126
100. Hunsen M, Azim A, Mang H, Wallner SR, Ronkvist A, Xie W, Gross RA (2007) Macromolecules 40:148
101. Dong H, Cao S-G, Li Z-Q, Han S-P, You D-L, Shen J-C (1999) J Pol Sci A Polym Chem 37:1265
102. Loeker FC, Duxbury CJ, Kumar R, Gao W, Gross RA, Howdle SM (2004) Macromolecules 37:2450
103. Takamoto T, Uyama H, Kobayashi S (2001) E-Polymers 004
104. Duxbury CJ, Wang W, de Geus M, Heise A, Howdle SM (2005) J Am Chem Soc 127:2384
105. Marcilla R, de Geus M, Mecerreyes D, Duxbury CJ, Koning CE, Heise A (2006) Eur Polym J 42:1215
106. Uyama H, Takamoto T, Kobayashi S (2002) Polym J 34:94
107. Duda A, Kowalski A, Libiszowski J, Penczek S (2005) Macromol Symp 224:71
108. Focarete ML, Scandola M, Kumar A, Gross RA (2001) J Pol Sci B Polym Phys 39:1721
109. Peeters J, Palmans ARA, Veld M, Scheijen F, Heise A, Meijer EW (2004) Biomacromolecules 5:1862
110. van As BAC, Thomassen P, Kalra B, Gross RA, Meijer EW, Palmans ARA, Heise A (2004) Macromolecules 37:8973
111. Eriksson M, Fogelstrom L, Hult K, Malmström E, Johansson M, Trey S, Martinelle M (2009) Biomacromolecules 10:3108
112. Simpson N, Takwa M, Hult K, Johansson M, Martinelle M, Malmström E (2008) Macromolecules 41:3613
113. de Geus M, Peters R, Koning CE, Heise A (2008) Biomacromolecules 9:752
114. Lee CS, Ru MT, Haake M, Dordick JS, Reimer JA, Clark DS (1998) Biotechnol Bioeng 57:686
115. Wittrup Larsen M, Zielinska DF, Martinelle M, Hidalgo A, Juhl Jensen L, Bornscheuer UT, Hult K (2010) ChemBioChem 11:796 doi:10.1002/cbic.200900743
116. Simpson N, Takwa M, Hult K, Johansson M, Martinelle M, Malmström E (2008) Macromolecules 41:3613
117. Takwa M, Hult K, Martinelle M (2008) Macromolecules 41:5230
118. Takwa M, Xiao Y, Simpson N, Malmström E, Hult K, Koning CE, Heise A, Martinelle M (2008) Biomacromolecules 9:704
119. Peeters JW, Palmans ARA, Meijer EW, Koning CE, Heise A (2005) Macromol Rapid Commun 26:684
120. Ceccorulli G, Scandola M, Kumar A, Kalra B, Gross RA (2005) Biomacromolecules 6:902
121. van Beek DJM, Gillissen MAJ, van As BAC, Palmans ARA, Sijbesma RP (2007) Macromolecules 40:6340
122. Magusin PCMM, Mezari B, van der Mee L, Palmans ARA, Meijer EW (2005) Macromol Symp 230:126

123. van der Mee L, Antens J, van de Kruijs B, Palmans ARA, Meijer EW (2006) J Polym Sci A Polym Chem 44:2166
124. Wisse E, Renken RAE, Roosma JR, Palmans ARA, Meijer EW (2007) Biomacromolecules 8:2739
125. Jiang Z, Azim H, Gross RA, Focarete ML, Scandola M (2007) Biomacromolecules 8:2262
126. Srivastava RK, Albertsson AC (2007) Macromolecules 40:4464
127. Srivastava RK, Albertsson AC (2006) Biomacromolecules 7:2531
128. Syrén P-O, Hult K (2010) ChemBioChem 11:802 doi:10.1002/cbic.200900758
129. van As BAC, van Buijtenen J, Heise A, Broxterman QB, Verzijl GKM, Palmans ARA, Meijer EW (2005) J Am Chem Soc 127:9964
130. From M, Adlercreutz P, Mattiasson B (1997) Biotechnol Lett 19:315
131. Torres C, Otero C (2001) Enzym Microb Technol 29:3
132. Wahlberg J, Persson PV, Olsson T, Hedenstroem E, Iversen T (2003) Biomacromolecules 4:1068
133. Hans M, Keul H, Möller M (2009) Macromol Biosci 9:239
134. Lassalle VL, Ferreira ML (2008) J Chem Technol Biotechnol 83:1493
135. Matsumura S, Mabuchi K, Toshima K (1997) Macromol Rapid Commun 18:477
136. Matsumura S, Mabuchi K, Toshima K (1998) Macromol Symp 130:285
137. Matsumura S, Tsukada K, Toshima K (1999) Int J Biol Macromol 25:161
138. Huijser S, Staal BBP, Huang J, Duchateau R, Koning CE (2006) Biomacromolecules 7:2465
139. Torres C, Otero C (1999) Enzym Microb Technol 25:745
140. Muller S, Uyama H, Kobayashi S (1999) Chem Lett 12:1317
141. Hilker I, Schaafsma AEJ, Peters RAH, Heise A, Nijenhuis AJ (2008) Eur Polym J 44:1441
142. Binns F, Taylor A, Roberts SM, Williams CF (1994) Br UK Pat Appl 2272904 (Baxenden Chemicals Ltd., UK)
143. Hu J, Gao W, Kulshrestha A, Gross RA (2006) Macromolecules 39:6789
144. Kulshrestha AS, Gao W, Fu H, Gross RA (2007) Biomacromolecules 8:1794
145. Kulshrestha AS, Gao W, Gross RA (2005) Macromolecules 38:3193
146. Kulshrestha AS, Sahoo B, Gao W, Fu H, Gross RA (2005) Macromolecules 38:3205
147. Bourg-Garros S, Razafindramboa N, Pavia AA (1998) Enzym Microb Technol 2:240
148. Flores MV, Sewalt JJW, Janssen AEM, van der Padt A (2000) Biotechnol Bioeng 67:364
149. Rodney RL, Stagno JL, Beckman EJ, Russell AJ (1999) Biotechnol Bioeng 62:259
150. van As BAC, van Buijtenen J, Mes T, Palmans ARA, Meijer EW (2007) Chemistry 13:832
151. Uyama H, Kuwabara M, Tsujimoto T, Kobayashi S (2003) Biomacromolecules 4:211
152. Mei Y, Kumar A, Gao W, Gross R, Kennedy SB, Washburn NR, Amis EJ, Elliott JT (2004) Biomaterials 25:4195
153. Iglesias LE, Fukuyama Y, Nonami H, Erra-Balsells R, Baldessari A (1999) Biotechnol Tech 13:923
154. Uyama H, Klegraf E, Wada S, Kobayashi S (2000) Chem Lett 7:800
155. Kulshrestha AS, Gao W, Fu HY, Gross RA (2007) Biomacromolecules 8:1794
156. Hu J, Gao W, Kulshrestha A, Gross RA (2008) ACS Symp Ser Polym Biocatal Biomater 999:275
157. Skaria S, Smet M, Frey H (2002) Macromol Rapid Commun 23:292
158. He F, Wang Y, Feng J, Zhuo R, Wang X (2003) Polymer 44:3215
159. Focarete ML, Gazzano M, Scandola M, Kumar A, Gross RA (2002) Macromolecules 35:8066
160. Liljeblad A, Kallio P, Vainio M, Niemi J, Kanerva LT (2010) Org Biomol Chem 8:886
161. Adam W, Groer P, Humpf H-U, Saha-Möller CR (2000) J Org Chem 65:4919
162. Forro E, Fulop F (2003) Org Lett 5:1209
163. Ditrich K (2008) Synthesis 2008(14):2283
164. van Rantwijk F, Sheldon RA (2004) Tetrahedron 60:501
165. Balkenhohl F, Ditrich K, Hauer B, Ladner W (1997) J Prakt Chem 339:381
166. Kricheldorf HR, Schwarz G (2007) J Macromol Sci A Pure Appl Chem 44:625
167. Kawada J, Lutke-Eversloh T, Steinbuchel A, Marchessault RH (2003) Biomacromolecules 4:1698
168. Lutke-Eversloh T, Bergander K, Luftmann H, Steinbuchel A (2001) Microbiology 147:11

169. Lutke-Eversloh T, Bergander K, Luftmann H, Steinbuchel A (2001) Biomacromolecules 2:1061
170. Lutke-Eversloh T, Kawada J, Marchessault RH, Steinbuchel A (2002) Biomacromolecules 3:159
171. Lutke-Eversloh T, Fischer A, Remminghorst U, Kawada J, Marchessault RH, Bogershausen A, Kalwei M, Eckert H, Reichelt R, Liu SJ, Steinbuchel A (2002) Nat Mater 1:236
172. Iglesias LE, Baldessari A, Gros EG (1996) Bioorg Med Chem Lett 6:853
173. Bianchi D, Cesti P (1990) J Org Chem 55:5657
174. Sproull KC, Bowman GT, Carta G, Gainer JL (1997) Biotechnol Progr 13:71

Adv Polym Sci (2010) 237: 79–113
DOI:10.1007/12_2010_74

Published online: 28 May 2010

Hydrolases in Polymer Chemistry: Chemoenzymatic Approaches to Polymeric Materials

Andreas Heise and Anja R.A. Palmans

Abstract Lipases show high activity in the polymerization of a range of monomers using ring-opening polymerization and polycondensation. The range of polymer structures from this enzymatic polymerization can be further increased by combination with chemical methods. This paper reviews the developments of the last 5–8 years in chemoenzymatic strategies towards polymeric materials. Special emphasis is on the synthesis of polymer architectures like block and graft copolymers and polymer networks. Moreover, the combination of chemical and enzymatic catalysis for the synthesis of unique chiral polymers is highlighted.

Keywords Block copolymers · Chirality · Dynamic kinetic resolution · Graft copolymer · Kinetic resolution · Polymer networks

Contents

A. Heise (✉)
School of Chemical Sciences, Dublin City University, Glasnevin, Dublin 9, Ireland
e-mail: andreas.heise@dcu.ie

A.R.A. Palmans (✉)
Laboratory of Molecular Science and Technology, Eindhoven University of Technology, PO Box 513, 5600 MB Eindhoven, The Netherlands
e-mail: a.palmans@tue.nl

Abbreviations

ATRP	Atom transfer radical polymerization
CALB	*Candida antarctica* Lipase B
CL	ε-Caprolactone
DXO	1,5-Dioxepane-2-one
FOMA	1*H*, 1*H*, 2*H*, 2*H*-Perfluorooctyl methacrylate
FRP	Free-radical polymerization
GMA	Glydidyl methacrylate
GPEC	Gradient polymer elution chromatography
HEA	Hydroxyethyl acrylate
HEMA	Hydroxyethyl methacrylate
LCCC	Liquid chromatography under critical conditions
MMA	Methyl methacrylate
NMP	Nitroxide-mediated polymerization
PEG	Poly(ethylene glycol)
PCL	Poly(ε-caprolactone)
PMCL	Poly(4-methyl-ε-caprolactone)
PGMA	Poly(glydidyl methacrylate)
PPDL	Poly(ω-pentadecalactone)
PTHF	Poly(tetrahydrofurane)
RAFT	Reversible addition–fragmentation chain transfer
ROP	Ring-opening polymerization
$scCO_2$	Supercritical carbon dioxide
SEC	Size-exclusion chromatography

1 Introduction

Chemists have developed numerous synthetic procedures to supply an ever-increasing number of structurally and functionally diverse materials for advanced application. Almost all of these procedures rely on efficient catalytic systems, which allow the synthesis to be carried out at an economical time-scale. Most catalysts are highly specific in terms of the chemical conversion they catalyze. Although this is an advantage for the catalytic selectivity, it limits the general applicability of a catalyst in multistep conversions. This can be overcome by the combination of several catalysts or methodologies, each specific for one conversion step. Depending on the compatibility of the individual reaction steps, this can either be done with or without isolation of the reaction intermediates (cascade reaction). This is not different for biocatalysts such as enzymes, which are tailored to very high substrate specificity in the natural environment. In many cases, the specificity of enzymes is retained in vitro and has led to improved (i.e., more efficient in terms of energy and raw materials) synthetic procedures (e.g., for pharmaceutical intermediates) in both stepwise and multienzyme cascade approaches [1]. However, not all chemical conversions can be achieved enzymatically, hence the full exploitation of multistep synthetic strategies will require the development of novel organic and biosynthetic methods, so-called chemoenzymatic procedures. Although a large number of chemoenzymatic procedures already exist in organic synthesis (both stepwise and cascade), this is a relatively new concept in polymer chemistry. Obviously, one reason is that enzymatic polymerization as such is a relatively new field and a basic understanding of its scope and limitations had to be developed first [2, 3]. On the other hand, polymer chemistry knows many examples in which two or more catalytic systems are combined to realize structures and materials not available from one technique alone [4]. Although enzymatic polymerizations, in particular those using hydrolases (lipases), are very successful for making a large variety of polymers by ring-opening polymerization (ROP) and polycondensation, a general shortcoming of lipase-catalyzed polymerizations is the lack of control over the molecular structure. This prohibits the synthesis of complex molecular architectures like block and graft copolymers, which are possible with controlled polymerization techniques.

Chemoenzymatic polymerizations have the potential to further increase macromolecular complexity by overcoming these limitations. Their combination with other polymerization techniques can give access to such structures. Depending on the mutual compatibility, multistep reactions as well as cascade reactions have been reported for the synthesis of polymer architectures and will be reviewed in the first part of this article. A unique feature of enzymes is their selectivity, such as regio-, chemo-, and in particular enantioselectivity. This offers opportunities to synthesize novel chiral polymers and polymer architectures when combined with chemical catalysis. This will be discussed in the second part of this article. Generally, we will focus on the developments of the last 5–8 years. Unless otherwise noted, the term enzyme or lipase in this chapter refers to *Candida antarctica* Lipase B (CALB) or Novozym 435 (CALB immobilized on macroporous resin).

2 Chemoenzymatic Synthesis of Polymer Architectures

2.1 Crosslinked Structures

Crosslinking of polymers is usually applied to stabilize the macroscopic morphology or shape of a material. In most cases, it results in insoluble polymeric materials, e.g., for polymeric coatings. In the chemoenzymatic strategies towards polymer networks, the enzymatic step is exclusively applied to synthesize the

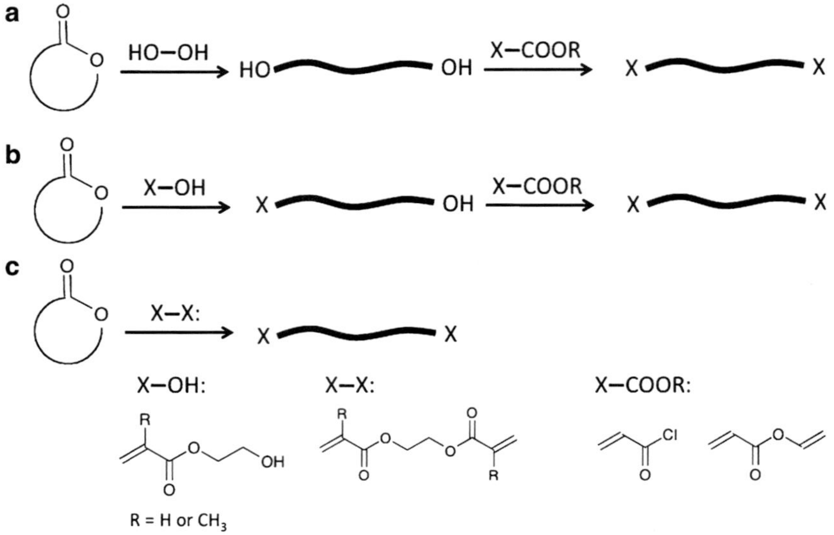

Fig. 1 Enzymatic routes (**a–c**) of telechelic diacrylates by enzymatic ROP for subsequent crosslinking. *X* denotes the polymerizable (meth)acrylate group

reactive polymer precursor, whereas the crosslinking step is achieved chemically. The reactive polymer precursors must have at least one or preferably two or more reactive groups capable of participation in a network-forming chemical reaction. The most common structures are polymers with acrylate endgroups obtained by the routes depicted in Fig. 1.

Following route A (Fig. 1), Yan Xiao et al. reported the chemoenzymatic synthesis of poly(ε-caprolactone) (PCL) and chiral poly(4-methyl-ε-caprolactone) (PMCL) microparticles [5]. The telechelic polymer diol precursors were obtained by enzymatic polymerization of the corresponding monomers in the presence of hexanediol. Enzymatic kinetic resolution polymerization directly yielded the (*R*)- and (*S*)-enriched chiral polymers. After acrylation using acryloylchloride, the chiral and nonchiral particles were obtained by crosslinking in an oil-in-water emulsion photopolymerization. Preliminary degradation experiments showed that the stereoselectivity of CALB is retained in the degradation of the chiral microparticles (Fig. 2).

A second route (route B in Fig. 1) relies on an initiation process with an (meth)acryl hydroxyl compound and is adopted from the chemical ROP of lactones. The controlled character of these polymerizations ensures a virtually quantitative initiation and thus incorporation of hydroxy-functional initiator (e.g., acrylate) into the polymer chain. However, this is not automatically the case for lipase-catalyzed ROP due to the different mechanism. The latter follows an activated monomer mechanism in which the lipase activates any carbonyl group of a carboxylic acid derivative present in the system. It has recently been shown that acrylation using hydroxy-functional acrylate initiators like hydroxy ethyl(meth)acrylate (HEMA or

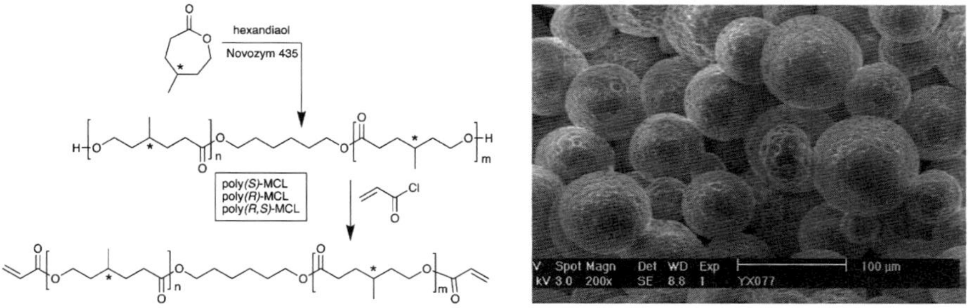

Fig. 2 Synthesis of acrylated chiral PMCL (*left*) and microparticles (*right*) obtained from acrylated polymers by oil-in-water emulsion photopolymerization [5]

HEA) according to route B is not straightforward. The lipase does not distinguish between the ester group of a monomer, a polyester chain, and the ester group present in HEA or HEMA. Consequently, the esters groups in HEA and HEMA will be activated as well and take part in transesterification processes, which results in a complex mixture of polymer endgroups when monomers like ε-caprolactone (CL) and ω-pentadecalactone (PDL) are used [6, 7]. The transesterification process occurs at moderate frequency at low monomer conversion but it becomes dominant at longer reaction times. Although this presents a restriction for the enzymatic synthesis of strictly mono-acrylated polymers from HEA and HEMA, fully dimethacrylated PCL and PPDL were obtained by combination with end-capping with vinyl methacrylate [6]. Taking advantage of the inevitable transesterification reactions, Takwa et al. reported an elegant single-step solvent-free enzymatic route to telechelic PPDL by conducting the enzymatic ROP in the presence of ethylene glycol di(meth)acrylate (route C in Fig. 1). Polymers with $>96\%$ diacrylate and dimethacrylate endgroups were obtained (Scheme 1) [8]. In the same paper, the authors introduced a synthetic route to telechelic dithioPPDL and thiol-acrylate functional PPDL. The thiol functional macromonomers were used together with other ene-monomers to give crosslinked PPDL thin films using a UV-induced photopolymerization [9]. In this first example of chemoenzymatic synthesis of coatings, semicrystalline networks were obtained owing to the high crystallinity of PPDL.

Various α-methylenemacrolides were enzymatically polymerized to polyesters having polymerizable methacrylic methylene groups in the main chain (Fig. 3, left). The free-radical polymerization of these materials produced crosslinked polymer gels [10, 12]. A different chemoenzymatic approach to crosslinked polymers was recently introduced by van der Meulen et al. for novel biomedical materials [11]. Unsaturated macrolactones like globalide and ambrettolide were polymerized by enzymatic ROP. The clear advantage of the enzymatic process is that polymerizations of macrolactones occur very fast as compared to the chemically catalyzed reactions [13]. Thermal crosslinking of the unsaturated polymers in the melt yielded insoluble and fully amorphous materials (Fig. 3, right).

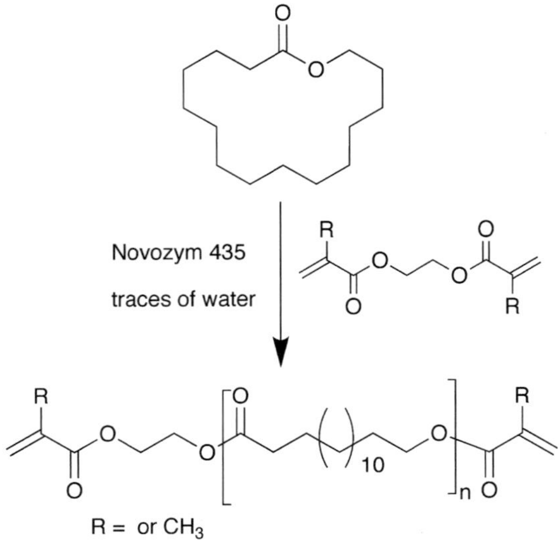

Scheme 1 Synthesis of diacrylated poly(pentadecalactone) [8]

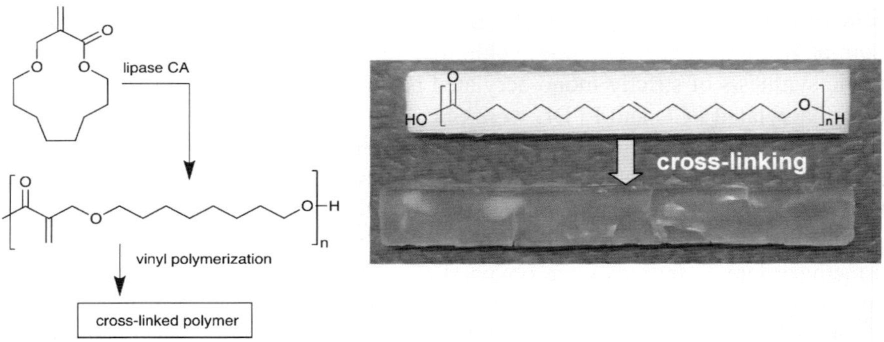

Fig. 3 Chemoenzymatic routes to crosslinked macrolactones. *Left*: enzymatic polymerization of α-methylenemacrolides and subsequent vinyl polymerization [10]. *Right*: Enzymatically synthesized highly crystalline poly(ambrettolide) before and after thermal crosslinking [11]

2.2 Copolymers

The synthesis of acrylates from sugars and other substrates has been applied in the early phases of enzymatic reactions and has already been reviewed [2, 14]. The method is attractive because the enzyme allows for mild and, in some examples, stereoselective acrylation. A recent example was published by Popescu et al. who took advantage of the frequent transesterification reactions and reported a route to highly functional linear copoly(meth)acrylates [15]. Methyl (meth)acrylate was mixed with various functional alcohols in the presence of Novozym 435. In situ transesterification leads to a mixture of functional (meth)acrylate monomers, which can be polymerized by free-radical polymerization without further workup to yield multifunctional poly(meth)acrylates.

2.3 Block Copolymers

Enzymatic polymerization has been combined with various chemical polymerizations for the synthesis of block copolymers. The choice of chemical polymerization generally depends on the applied strategy for the block copolymer synthesis. These can be divided into three main approaches, as shown in Fig. 4 for the example of enzymatic ROP. It has to be noted that some of these strategies have also been applied for enzymatic polycondensations.

Initial reports on chemoenzymatic block copolymer synthesis focus on the enzymatic macroinitiation from chemically obtained hydroxy-functional polymers (route A in Fig. 4). The first report on enzymatic macroinitiation was published by Kumar et al., who used anionically synthesized hydroxy-functional polybutadiene of various molecular weights ranging from 2600 to 19,000 Da (Fig. 5) [16]. In a systematic study, the authors investigated the efficiency of the macroinitiation of CL and PDL by Novozym 435 as a function of the polybutadiene macroinitiator. The reaction profile showed that polybutadiene consumption steadily increased with the reaction

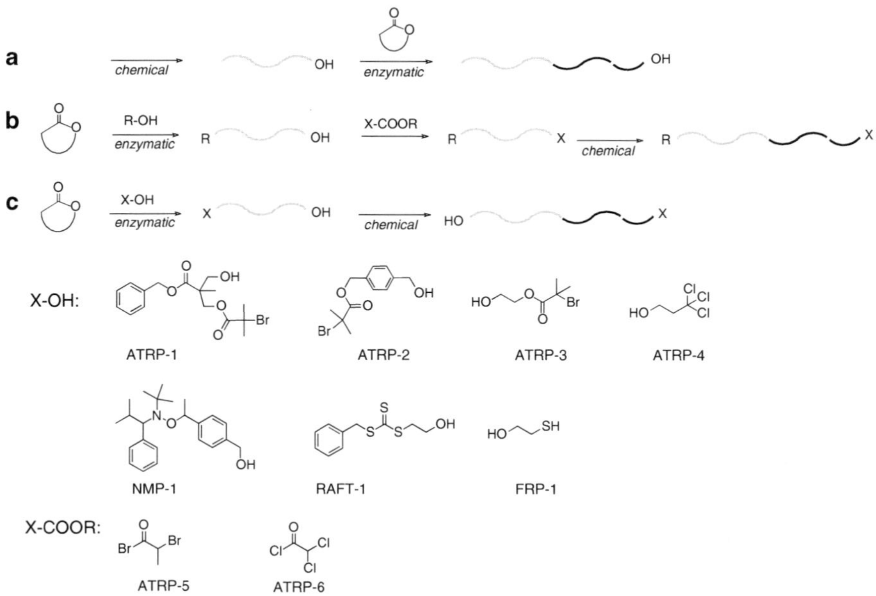

Fig. 4 Chemoenzymatic strategies (**a–c**) for the synthesis of block copolymers employing enzymatic ROP and radical polymerization techniques

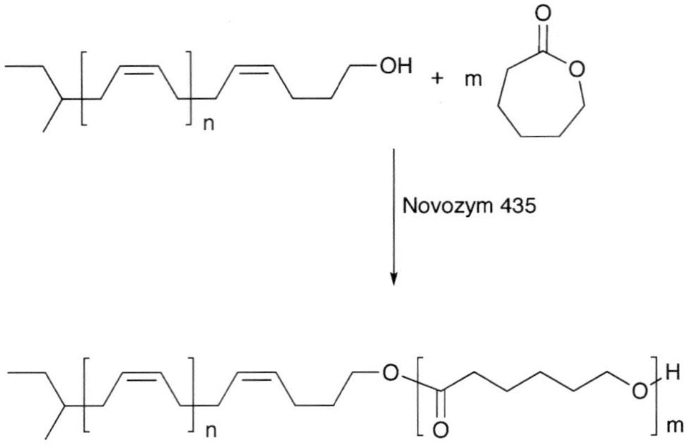

Fig. 5 Synthesis of poly(butadiene-*b*-caprolactone) from hydroxy-functional polybutadiene [16]

time, according to incorporation by initiation and transesterification. The latter is due to the fact that the enzyme activates both the carbonyl bond of the monomer (CL) and the PCL esters. Consequently, the nucleophile (initiator) will be incorporated by both mechanisms, i.e., initiation and transesterification. It was found that all polybutadiene macroinitiators formed block copolymers irrespective of the molecular weight. Initiation efficiency was >80%, and the presence of water-initiated chains was less than 30%.

Several authors reported the use of hydroxy-functional poly(ethylene glycol) (PEG) as macroinitiator for enzymatic ROP. Panova and Kaplan conducted a detailed mechanistic and kinetic study of the interplay between monomer conversion, chain initiation, and chain propagation of the enzymatic ROP of CL in the presence of monohydroxy-functional methoxy-PEG [17]. The results confirmed the dynamic character of the reaction and the presence of a product mixture containing unreacted macroinitiator, water-initiated PCL, cyclic PCL, and the corresponding block copolymer. At otherwise identical reaction conditions, the amount of reacted monohydroxy-functional methoxy-PEG was concentration-dependent, being 35% at higher concentrations and up to 84% at lower concentrations for identical reaction times. Moreover, block copolymers from mono- and difunctionalized PEG were obtained by Feng He et al. by the enzymatic ROP of CL [18], and by Srivastava and Albertsson with CL and 1,5-dioxepane-2-one (DXO) [19]. Kaihara et al. reported the enzymatic ROP of trimethylene carbonate from a copolymer of PEG and a cyclic acetale with two terminal hydroxyl groups [20]. The resulting amphiphilic triblock copolymer was used to form micellar structures, and pH-dependent drug release was successfully shown.

While all previous examples employ enzymatic ROP, there are two reports on block copolymer synthesis employing enzymatic polycondensation. The first one was published by Sharma et al. and describes the synthesis and solid-state properties of polyesteramides with poly(dimethylsiloxane) (PDMS) blocks [21]. The polycondensation was carried out with various ratios of dimethyl adipate, octanediol, and diamine-functionalized PDMS ($M_w = 875$). Depending on the composition of the polymer, the physical properties varied from hard solid to sticky materials. The second report, by Niu et al., describes the synthesis of multiblock copolymers of poly(tetrahydrofurane) (PTHF) and PEG [22]. In a two-step procedure, first the hydroxyl groups of PTHF diol ($M_n = 1000$ Da) were enzymatically reacted with diethyl malonate. In the second step, an enzymatic polycondensation with PEG diols of various molecular weights was carried out to give copolymers with molecular weights of 13,000–18,000 Da.

In the majority of reports on the chemoenzymatic synthesis of block copolymers, the enzymatic polymerization is conducted first (routes B and C in Fig. 4). Mostly, controlled polymerization techniques like atom transfer radical polymerization (ATRP), nitroxide-mediated polymerization (NMP), or reversible addition–fragmentation chain transfer (RAFT), but also free-radical and controlled ROP, are applied as the nonenzymatic polymerization techniques. The controlled polymerization techniques offer the advantage of high initiation efficiency and control over the molecular weight, which generally results in well-defined block copolymers. This method requires the end-functionalization of the enzymatic block with an initiator moiety capable of selectively initiating the subsequent polymerization technique. A very elegant way to achieve that is by using a dual or bifunctional initiator (route C in Fig. 4). These initiators have a functional group capable of initiating the enzymatic polymerization (hydroxyl) and a group for the second polymerization. This approach has the advantage that any polymer analogous endgroup modification is avoided. On the other hand, for a high block copolymer yield it is important to ensure a high degree of incorporation of the dual initiator into the enzymatic block. Considering that nucleophiles such as the applied dual initiators are incorporated by initiation and transesterification in a similar way to the HEA and HEMA initiators, controlling the endgroup fidelity in this process becomes crucial. In the following paragraphs, the chemoenzymatic synthesis of block copolymer according to routes B and C is reviewed according to the applied chemical method.

ATRP is the method most widely applied in this synthetic strategy. The initial contribution was been made by Meyer et al., who reported the combination of enzymatic ROP and ATRP using the dual initiator ATRP-1 (Fig. 4) [23]. Macroinitiation of styrene from the ATRP-initiator-functionalized PCL yielded the block copolymer. In a follow-up paper, the same authors further investigated the influence of the reaction conditions and the initiator structure on the chemoenzymatic synthesis of block copolymer by this combination of polymerization techniques [24]. Special attention was given to the efficiency of the dual initiator in the enzymatic ROP of CL by comparison of the initiators ATRP-1, ATRP-2, and ATRP-3. It was found that the latter two are incorporated relatively fast (>90% at 15 min reaction time), whereas ATRP-1 was incorporated at a very slow rate (40% at 60 min reaction time). Liquid chromatography under critical conditions (LCCC) further confirmed that ATRP-1 is a poor initiator and is predominantly incorporated by transesterification [25]. Moreover, de Geus et al. investigated the possibility of conducting both polymerizations in the presence of the dual initiator, ATRP catalyst, CL, and methacrylates as a one-pot cascade reaction (Fig. 6) [26]. The study revealed that ATRP catalysts

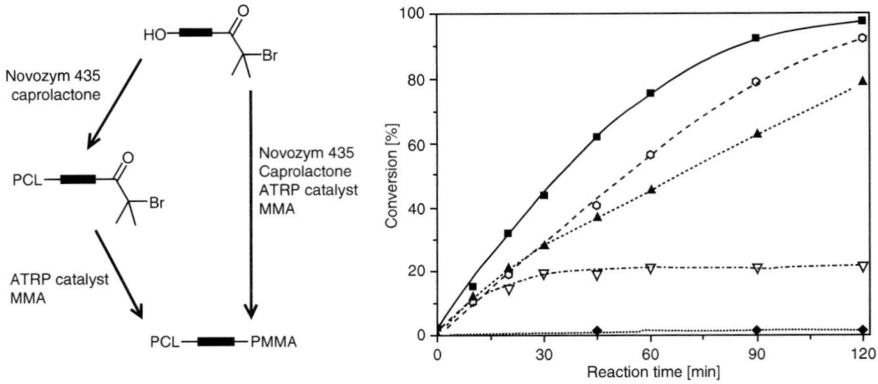

Fig. 6 *Left*: Strategy for consecutive chemoenzymatic and simultaneous one-pot block copolymer synthesis combining enzymatic ROP and ATRP. *Right*: Influence of ATRP–catalyst system on the conversion of CL in the enzymatic ROP of MMA at 60 °C using ATRP-3 as initiator: *filled squares* reaction in absence of ATRP-catalyst; *open circles* CuBr/PMDETA (1:1:1 ratio with respect to initiator); *filled triangles* CuBr/dNbpy (1:2.1:1 ratio with respect to initiator); *open inverted triangles* CuBr (1:1 ratio with respect to initiator); *filled diamonds* $CuBr_2$ (1:1 ratio to initiator). CL conversion was determined with ^{1}H-NMR [26]

like nickel have a strong inhibiting effect on the enzyme, whereas typical ATRP copper catalysts were tolerated by the enzyme (Fig. 6). The chemoenzymatic cascade reaction was most successful when the ATRP components were added after the enzymatic process. High block copolymer yields were obtained in that approach, whereas homopolymer impurities were observed when all components were present from the beginning.

By taking advantage of the simultaneous enzyme inhibition by nickel, the nickel-catalyzed ATRP, and the stereoselectivity of the enzyme, Peters et al. obtained chiral block copolymers by this method from 4-methyl-ε-caprolactone (4-MeCL) by [27]. The polymerization of racemic 4-MeCL showed good enantioselectivity and produced a chiral macroinitiator with ATRP endgroup by selectively polymerizing only the (*S*)-4-MeCL. Macroinitiation was then started by adding the nickel catalyst and methyl methacrylate (MMA) to the reaction mixture, which simultaneously inhibited the enzyme and activated the ATRP process. Chiral poly[MMA-*b*-(*S*)-4-MeCL] was successfully obtained in this synthesis.

Howdle and coworkers reported that a one-pot, simultaneous synthesis of block copolymers by enzymatic ROP and ATRP employing initiator ATRP-3, CL, and MMA is possible in supercritical CO_2 ($scCO_2$) [28]. The authors could show that the CL acts as a $scCO_2$ cosolvent and was crucial for the radical polymerization to remain homogeneous and controlled. The unique ability of $scCO_2$ to solubilize highly fluorinated species was utilized by extending this methodology to the synthesis of novel copolymers consisting of a semifluorinated block of poly(1*H*, 1*H*, 2*H*, 2*H*-perfluorooctyl methacrylate) (PFOMA) and PCL [29]. Block copolymers were successfully synthesized by a two-step process based on the sequential monomer

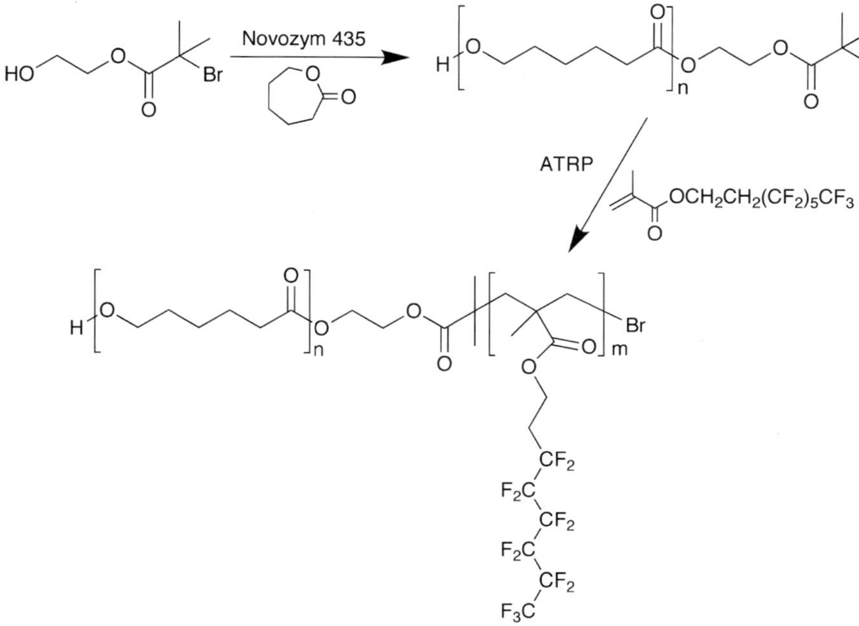

Scheme 2 Chemoenzymatic synthesis of block copolymers consisting of a semifluorinated block of poly(1*H*, 1*H*, 2*H*, 2*H*-perfluorooctyl metharylate) and PCL [29]

addition (Scheme 2). Parallel experiments in conventional solvents did not yield any block copolymers due to the limited solubility of FOMA in these solvents. These and other enzymatic polymerization procedures in $scCO_2$ have recently been reviewed [30].

Sha et al. applied the commercially available dual initiator ATRP-4 for the chemoenzymatic synthesis of block copolymers. In a first series of publications, the group reported the successful synthesis of a block copolymer comprising PCL and polystyrene (PS) blocks [31, 32]. This concept was then further applied for the chemoenzymatic synthesis of amphiphilic block copolymers by macroinitiation of glycidyl methacrylate (GMA) from the ATRP functional PCL [33]. This procedure yielded well-defined block copolymers, which formed micelles in aqueous solution. Sha et al. were also the first to apply the dual enzyme/ATRP initiator concept to an enzymatic polycondensation of 10-hydroxydecanoic acid [34]. This concept was then extended to the ATRP of GMA and the formation of vesicles from the corresponding block copolymer [35].

The second strategy for the chemoenzymatic synthesis of block copolymers from enzymatic macroinitiators employs an individual modification step of the enzymatic block with an initiator for the chemical polymerization (route B in Fig. 4). This strategy has the advantage that it does not depend on a high incorporation rate of the dual initiator. On the other hand, quantitative end-functionalization becomes more

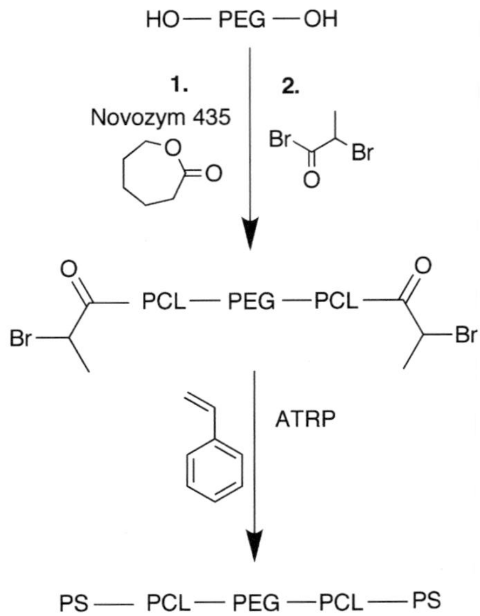

Fig. 7 Chemoenzymatic synthesis of the pentablock copolymer PS-*b*-PCL-*b*-PEG-*b*-PCL-*b*-PS [39]

difficult with increasing molecular weight of the polymer. Initial work following this approach was carried out by Sha et al., who reported the quantitative functionalization of enzymatically obtained PCL with α-bromopropionyl bromide (ATRP-5) [36]. The subsequent ATRP of styrene yielded the block copolymer. Based on this successful strategy, GMA was also polymerized from the PCL macroinitiator to yield amphiphilic bock copolymers, which assemble into nanoscale micelles [37]. The same author also extended this concept to the synthesis of hydroxy-functional polymer macroinitiators by enzymatic polycondensation of 10-decanoic acid [38]. Pentablock copolymers were also obtained by combining this approach with the enzymatic macroinitiation of CL from PEG diol (Fig. 7) [39]. The obtained PCL diol was modified with α-bromopropionyl to yield a telechelic ATRP macroinitiator. After ATRP of styrene, the obtained material, namely PS-*b*-PCL-*b*-PEG-*b*-PCL-*b*-PS, revealed interesting self-assembly behavior in water, forming spherical micelles, rod-like micelles, vesicles, lamellae, and large compound micelles depending on the block copolymer composition and concentration.

A series of interesting block copolymer architectures has also been prepared by Zhang et al. In a first paper, the synthesis of H-shaped triblock copolymers was demonstrated from enzymatically obtained PCL diol after end-functionalization with a difunctional ATRP initiator [40]. This allowed the growth of two PS chains from each end of the telechelic PCL. When methanol instead of glycol was used as the initiator in the initial enzymatic CL polymerization, a PCL with one hydroxyl endgroup was obtained. Functionalization of this endgroup with the difunctional ATRP initiator and subsequent ATRP of styrene or GMA resulted in Y-shaped polymers (Scheme 3) [41, 42].

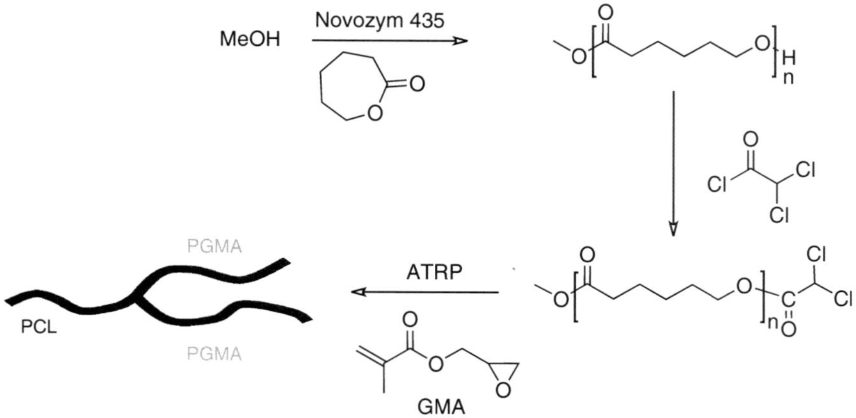

Scheme 3 Synthesis of Y-shaped polymers combining ATRP and enzymatic ROP [41, 42]

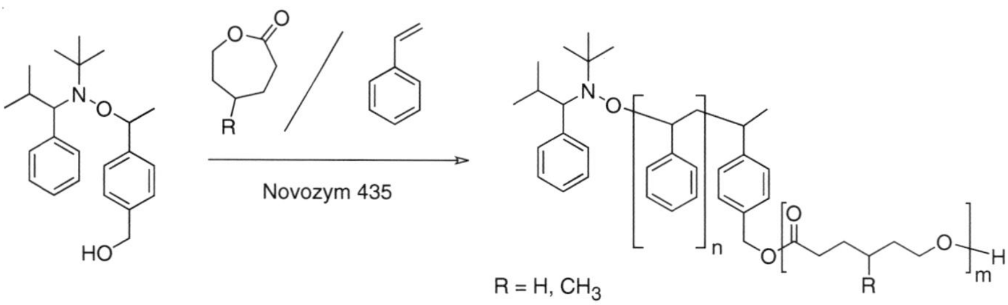

Scheme 4 One-pot chemoenzymatic cascade polymerization combining enzymatic ROP and NMP for the synthesis of (chiral) block copolymer [43]

Apart from ATRP, the concept of dual initiation was also applied to other (controlled) polymerization techniques. Nitroxide-mediated living free radical polymerization (LFRP) is one example reported by van As et al. and has the advantage that no further metal catalyst is required [43]. Employing initiator NMP-1, a PCL macroinitiator was obtained and subsequent polymerization of styrene produced a block copolymer (Scheme 4). With this system, it was for the first time possible to successfully conduct a one-pot chemoenzymatic cascade polymerization from a mixture containing NMP-1, CL, and styrene. Since the activation temperature of NMP is around 100 °C, no radical polymerization will occur at the reaction temperature of the enzymatic ROP. The two reactions could thus be thermally separated by first carrying out the enzymatic polymerization at low temperature and then raising the temperature to around 100 °C to initiate the NMP. Moreover, it was shown that this approach is compatible with the stereoselective polymerization of 4-MeCL for the synthesis of chiral block copolymers.

Similarly successful was the combination of enzymatic ROP with RAFT in $scCO_2$, as reported by Howdle [44]. Employing the initiator RAFT-1, a block copolymer comprising PCL and PS was obtained and the block structure confirmed by GPEC. Like NMP, RAFT has the advantage that it does not require a metal catalyst and, when coupled with the green solvent CO_2, this process is environmentally desirable.

Also, Kerep and Ritter reported a radical chain transfer agent as a dual initiator, FRP-1 [45]. The first step builds on the fact that hydroxyl groups are much better nucleophiles in enzymatic ROP than thiols. Due to the chemoselectivity of the enzyme, PCLs with predominantly thiol endgroups were obtained, which were subsequently used as macroinitiator for styrene. The authors report that the reaction yield can be further increased by microwave irradiation. Although thiols provide less control over the radical polymerization than RAFT agents, the subsequent radical polymerization successfully leads to the synthesis of PCL-*b*-PS.

Block copolymers were also obtained by combining enzymatic with chemically catalyzed ROP. An approach for the synthesis of multiarm heteroblock copolymers comprising PCL and PLA was reported by Gross et al. [46]. Multifunctional 1-ethyl glucopyranoside was first used as an initiator for the enzymatic ROP of CL. Due to the selectivity of the enzyme from *Pseudomonas cepacia*, only the one primary hydroxyl group initiated the polymerization, while the three secondary alcohols remained unfunctionalized. By subsequent tin-catalyzed ROP of lactide, the secondary alcohols were activated and three PLA arms were grown from the PCL macroinitiator. Very recently, Yan Xiao et al. reported the first example of the combination of enzymatic and chemical carbene-catalyzed ROP for the formation of block copolymers (Scheme 5) [47]. In their approach, the authors took advantage of the fact that CALB has a high catalytic activity for lactones but not for lactides (LLA), whereas it is vice versa for carbenes. In the synthetic strategy, the enzymatic polymerization of CL was conducted first. Addition of LLA and carbene to the reaction mixture led to the macroinitiation directly from the hydroxyl endgroup of PCL, allowing the synthesis of PCL-*b*-PLLA block copolymers.

A synthetic approach to poly(ester-urethanes) was recently published by Xue et al. [48]. Glycerol was employed as a trifunctional initiator in the enzymatic ROP of CL. The three-arm PCL triol was then reacted with methylene-diphenyl diisocyanate (MDI) and hexanediol to yield a three-arm PCL-based poly(ester-urethane) with shape-memory properties.

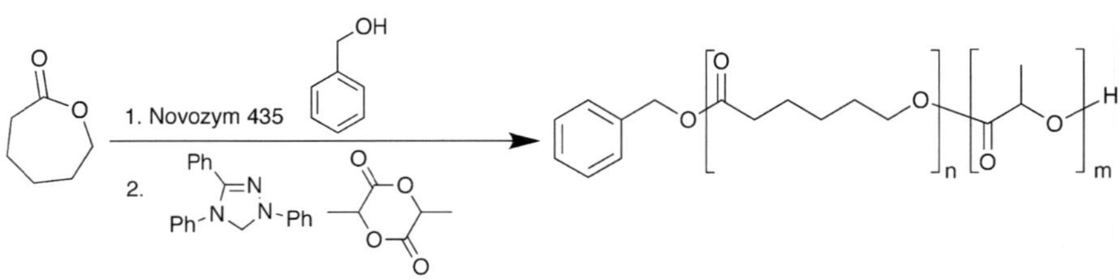

Scheme 5 Combination of enzymatic and chemical carbene-catalyzed ROP for the metal-free formation of block copolymers [47]

2.4 Graft Copolymers

Graft copolymers were prepared by both classical strategies, i.e., from enzymatically obtained macromonomers using subsequent chemical polymerization and by enzymatic grafting from hydroxy-functional polymers.

Kalra et al. studied the synthesis of PPDL graft copolymers following route A as shown in Scheme 6. The macromonomers were obtained by the enzymatic ROP of PDL from HEMA and PEGMA [49]. In a comparative study, Novozym 435 was found to be the most active biocatalyst for this reaction step. Subsequently, graft copolymers were obtained by free-radical polymerization of the macromonomers. A similar approach was published by Srivastava for the HEMA-initiated enzymatic ROP of CL and subsequent free-radical polymerization [50].

A significant contribution to the enzymatic grafting from multifunctional macroinitiators was made by Keul et al. using well-defined polyglycidols as promising biomedical materials (Scheme 6, route B) [51]. In an initial paper, the authors compared the chemical and enzymatic ROP of CL from linear and star-shaped polyglycidol obtained from anionic ROP [52]. It was found that the zinc-catalyzed ROP of CL from the multifunctional polyglycidol resulted in a quantitative initiation efficiency, whereas enzymatic ROP PCL was only grafted to 15–20% of the hydroxyl groups. This leads to different polymer architectures, i.e., the fully grafted polymer resembles a core–shell structure with hydrophilic polyether core and hydrophobic PCL shell. The enzymatically obtained partly grafted polymer, on the other hand, has a hydrophilic polyether head with a hydrophobic PCL tail. The difference in the initiation efficiency is due to the different polymerization mechanisms. Whereas the chemical ROP is endgroup-activated and controlled, the enzymatic ROP is monomer-activated. In the latter case, the steric constraints in the nucleophilic attack of the hydroxyl groups on the enzyme-activated monomer prevents the reactions of all hydroxyl groups. This was also confirmed by a study of Duxbury et al. of the enzymatic ROP from PS containing 10% of 4-vinylbenzyl alcohol [53]. Although quantitative enzymatic modification of the hydroxyl groups with the small vinyl acetate was observed, the grafting of PCL produced only a 50–60% grafting efficiency. Moreover, the results suggested that the grafting action is a combination of grafting from and grafting onto by transesterification of PCL.

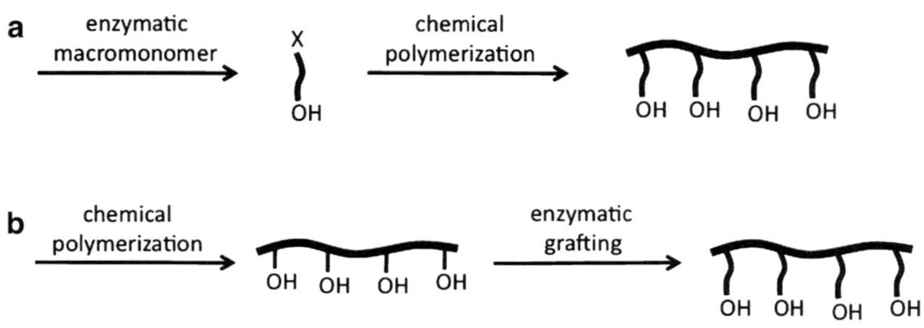

Scheme 6 Enzymatic routes (**a** and **b**) towards graft copolymers

The incomplete enzymatic grafting from polyalcohols opens opportunities for the synthesis of unique structures, as shown in several examples by Hans and coworkers. For example, heterografted molecular bottle brushes were synthesized starting from PCL-grafted polyglicydol [54]. The grafting yield in this step was about 50%. Selective acetylation of the hydroxyl groups at the PCL graft ends was achieved via enzymatic reaction with vinyl acetate without acetylation of the remaining hydroxyl groups at the polyglycidol backbone. The latter were quantitatively converted to ATRP initiator groups, and initiation of MMA *n*-butyl acrylate yielded well-defined heterografted polymers. Alternatively, the remaining hydroxyl groups on the polyglycidol backbone were used to initiate chemically catalyzed ROP of lactide to produce a heterografted polymer comprising PCL and PLA grafts [55]. When the endgroups of the PCL were capped with acrylates, the resulting materials could be formulated into UV-crosslinked microspheres (Fig. 8).

The first synthesis of graft copolymers in $scCO_2$ was reported by Villarroya et al. from a polyHEMA–PMMA random copolymer by the combination of ATRP and enzymatic ROP of CL [56]. The authors first synthesized the random copolymer containing 13 and 30% HEMA by ATRP in $scCO_2$. Although PCL grafting was confirmed, not all hydroxyl groups participated in the grafting reaction (33%). Similar to the situation in conventional solvents, steric effects were thought to reduce the grafting efficiency. The results were similar when both the ATRP reaction and enzymatic grafting were carried out simultaneously in one-pot, i.e., all components for the radical and the enzymatic reaction were present at the same time. This approach resulted in about 40% functionalization of the hydroxyl groups. In another study, the authors confirmed that the limitation in grafting efficiency was due to two reasons: (1) the poly(MMA-*co*-HEMA) probably has a blocky structure due to

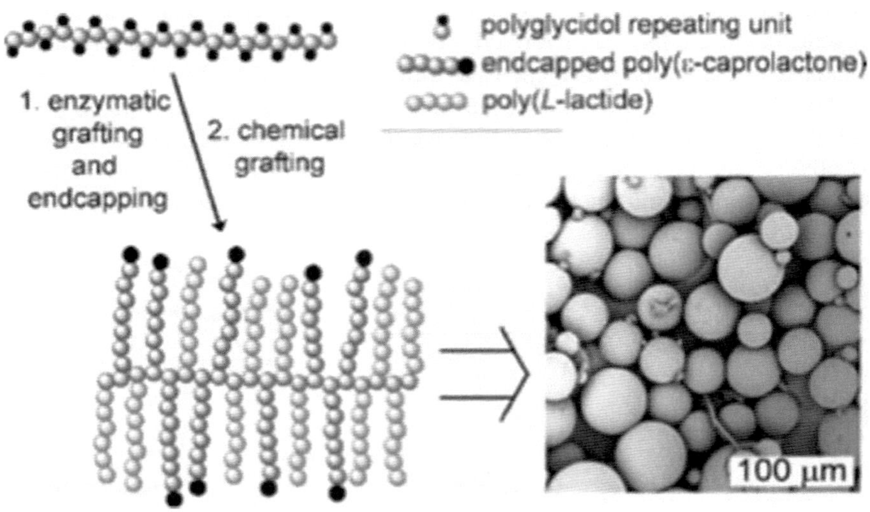

Fig. 8 Chemoenzymatic synthesis of heterografted polymers comprising PCL and PLA grafts on a polyglycidol backbone. After acrylation, these polymers can be formulated into crosslinked microspheres [55]

the different reactivity ratios of the monomers in the radical polymerization; and (2) the hydroxyl groups are too close to the polymer backbone. Consequently, the grafting efficiency was significantly improved when highly randomized poly(MMA-*co*-HEMA) from starved-feed polymerization was used (80%), and when a PEG spacer was introduced between the polymer backbone and the hydroxyl group (100%) [57].

2.5 Branched Structures

The only report on chemoenzymatic synthesis of branched polymers is from Peeters et al. [58]. Heterotelechelic PCL macroinimer was synthesized in a one-pot enzymatic procedure by using 2-hydroxyethyl α-bromoisobutyrate as a bifunctional initiator. A polymerizable endgroup was introduced by subsequent in situ enzymatic acrylation with vinyl acrylate. Synthesis of branched polymers by self-condensing ATRP of the macroinimers was successfully conducted with and without the addition of MMA as a comonomer.

3 Chemoenzymatic Approaches to Chiral (Co)polymers

3.1 Enantioselectivity Issues in Enzyme-Catalyzed Reactions

Introducing chirality into polymers has distinctive advantages over the use of nonchiral or atactic polymers because it adds a higher level of complexity, allowing for the formation of hierarchically organized materials. This may have benefits in high-end applications such as nanostructured materials, biomaterials, and electronic materials. Synthetically, chiral polymers are typically accessed by two methods. Firstly, optically active monomers – often obtained from natural sources – are polymerized to afford chiral polymers. Secondly, chiral catalysts are applied that induce a preferred helicity or tacticity into the polymer backbone or activate preferably one of the enantiomers [59–64].

Polymers derived from natural sources such as proteins, DNA, and polyhydroxyalkanoates are optically pure, making the biocatalysts responsible for their synthesis highly appealing for the preparation of chiral synthetic polymers. In recent years, enzymes have been explored successfully as catalysts for the preparation of polymers from natural or synthetic monomers. Moreover, the extraordinary enantioselectivity of lipases is exploited on an industrial scale for kinetic resolutions of secondary alcohols and amines, affording chiral intermediates for the pharmaceutical and agrochemical industry. It is therefore not surprising that more recent research has focused on the use of lipases for synthesis of chiral polymers from racemic monomers.

With the increased use of enzymes in polymer chemistry, the enzymology terminology to describe the reaction kinetics and the enantioselectivity of a reaction has become more and more common in polymer literature. The parameter of choice to describe the enantioselectivity of an enzyme-catalyzed kinetic resolution is the enantiomeric ratio E. The enantiomeric ratio is defined as the ratio of the specificity constants for the two enantiomers, (R) and (S) (1):

$$E = (k_{\text{cat}}/K_{\text{M}})_{\text{R}}/(k_{\text{cat}}/K_{\text{M}})_{\text{S}} \tag{1}$$

where k_{cat} is the rate constant, K_{M} the Michaelis–Menten constant, and $k_{\text{cat}}/K_{\text{M}}$ the specificity constant. Sih et al. developed (1) in terms of the enantiomeric excess of product or substrate and the conversion, both for reversible and irreversible reactions [65]. For an irreversible reaction, which is preferably the case in a polymerization reaction, (2) can be used to calculate E from either the substrate enantiomeric excess (ee_{S}) or the product enantiomeric excess (ee_{p}) and the conversion c:

$$E = \frac{\ln[1 - c(1 + ee_{\text{p}})]}{\ln[1 - c(1 - ee_{\text{p}})]} = \frac{\ln[(1 - c)(1 - ee_{\text{S}})]}{\ln[(1 - c)(1 + ee_{\text{S}})]}. \tag{2}$$

Because calculation of E on the basis of one conversion and ee measurement is highly unreliable, curve fitting should be employed using as many data points as possible. Figure 9 shows the conversion versus ee_{S} curves for Es of 3, 5, 10, and 100. At 50% conversion, the values of ee_{S} differ the most, indicating that measurements at around 50% conversion provide more valuable data than measurements at low or high conversion levels.

In order to get good results in a kinetic resolution, E must be high, preferably well above 30 [66]. However, in organic media the enantioselectivity of an enzyme can depend strongly on parameters such as temperature and solvent, so medium engineering is often a fast and highly effective tool to increase E.

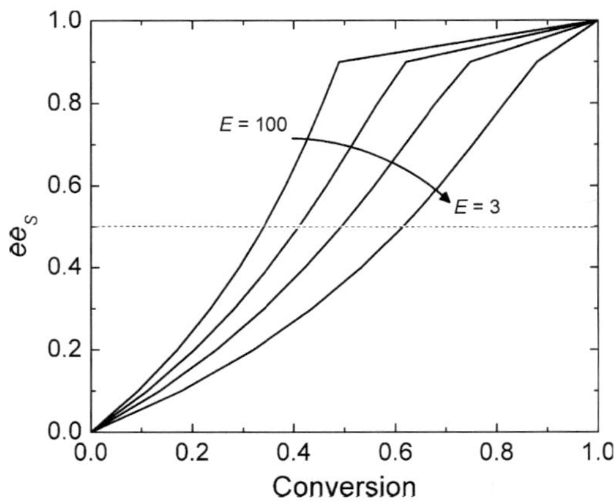

Fig. 9 Theoretical conversion versus ee_{S} curves for different Es

3.2 *Reaction Mechanism and Enantioselectivity of Lipases*

Lipases belong to the subclass of α/β-hydrolases and their structure and reaction mechanism are well understood. All lipases possess an identical catalytic triad consisting of an aspartate or glutamate, a histidine, and a nucleophilic serine residue [67]. The reaction mechanism of CALB is briefly discussed as a typical example of lipase catalysis (Scheme 7).

The catalytic triad of CALB consists of Asp187, His224, and Ser105, while the oxyanion hole is formed by the backbone amide protons of Thr40 and Gln106 and the side-chain of Thr40. First, a substrate reversibly complexes to the free enzyme (Scheme 7, top left), thereby forming a Michaelis–Menten complex. After correct positioning of the substrate, a nucleophilic attack of Ser105 onto the substrate carbonyl group occurs and a first tetrahedral intermediate is formed (Scheme 7, top right). In this tetrahedral intermediate, the negative charge on the former substrate carbonyl oxygen is stabilized by three hydrogen-bond interactions with the oxyanion hole, whereas the positive charge on His224 is stabilized by interaction with Asp187. Subsequently, proton transfer from His224 to the substrate alkyl oxygen takes place and the alcohol part of the residue is liberated from the enzyme. As a result, a covalently bound acyl-enzyme intermediate is formed at the end of the acylation step (Scheme 7, bottom right). Subsequently, the acyl-enzyme intermediate is deacylated by an incoming nucleophile R″NuH, which is generally water, an alcohol, or an amine. A second tetrahedral intermediate is formed by attack of the nucleophile onto the acyl-enzyme carbonyl group (Scheme 7, bottom left). In this process, the proton is transferred from the nucleophile to the His224 residue and the

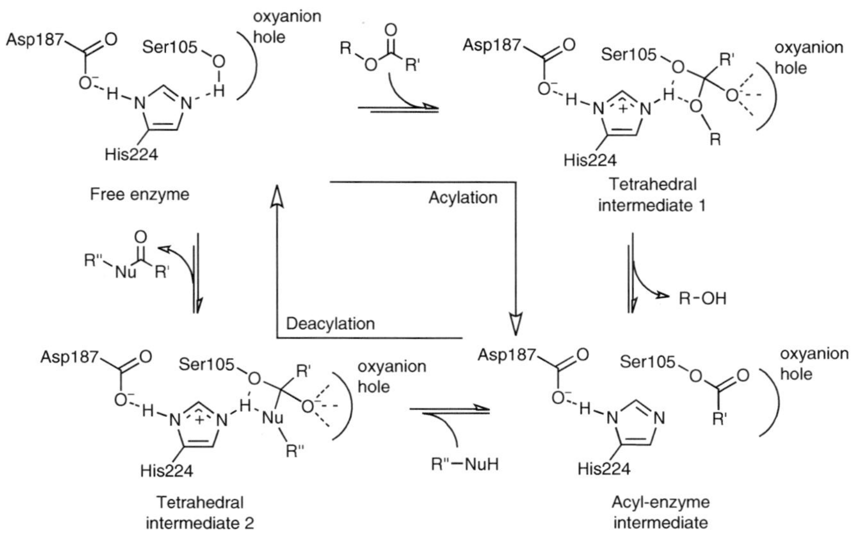

Scheme 7 Catalytic mechanism of CALB showing an acylation and deacylation step and the formation of a covalently bound acyl-enzyme intermediate (*bottom right*)

positive and negative charges are effectively stabilized by Asp187 and the oxyanion hole, respectively. Then, the proton is transferred from the His224 residue to the Ser105 alkyl oxygen while restoring the carbonyl bond of the bound substrate. As a result, a weakly bound enzyme–product complex is formed and the free enzyme species is regenerated after release of the reaction product.

3.2.1 Enantioselectivity of Lipases Towards the Nucleophile

The enantioselectivity of lipases towards secondary alcohols and amines is well-documented in the literature [68–71]. In general, lipases display a strong (*R*)-selectivity towards secondary alcohols. In the case of CALB, for example, an E of 10^6 has been reported in the kinetic resolution of (*rac*)-1-phenylethanol [72]. The active site of CALB contains a small cavity called the stereospecificity pocket, which can hold a methyl- or ethyl-sized substituent. The smaller substituent of the secondary alcohol can be placed in this stereospecificity pocket during reaction, while the larger substituent resides in the larger cavity. The spatial orientation of these cavities implies that (*R*)-secondary alcohols are highly preferred; the docking of an (*S*)-secondary alcohol leads to significant steric hindrance. This behavior was empirically recognized by Kazlauskas [73] and was later confirmed on a molecular level by Uppenberg et al. [74]. Similarly, the (*R*)-enantiomer in primary amines next to a chiral center is also the faster reacting enantiomer.

3.2.2 Enantioselectivity of Lipases Towards the Acyl Donor

Although many publications have covered the enantioselectivity of lipases in the deacylation step, their enantioselectivity in the acylation step (i.e., towards the acyl donor) has received much less attention. Generally, the selectivity of lipases towards racemic esters or acids is low to moderate [75–77]. Directed evolution and site-directed mutagenesis lead to a significant increase in the selectivity of the wild-type enzymes [78–80]. However, the enantiomeric ratios attained are still well below those typically obtained in kinetic resolutions of secondary alcohols.

3.2.3 Enantioselectivity in the Ring-Opening of Substituted Lactones

Lactones are ideal substrates for preparing polymers because they incorporate both the acyl donor as well as the nucleophile (after ring-opening) in one molecule. Introducing a substituent on the lactone ring inevitably generates a chiral center. Since the enantioselectivity of an enzyme can be determined both in the acylation and in the deacylation step, the question is how the position and the size of a substituent on the lactone ring affects the selectivity. In addition, the size of the lactone ring may play an important role because the ring conformation of the ester bond in the lactone ring changes from *cisoid* (ring size <8) to *transoid* (ring size >8) [81].

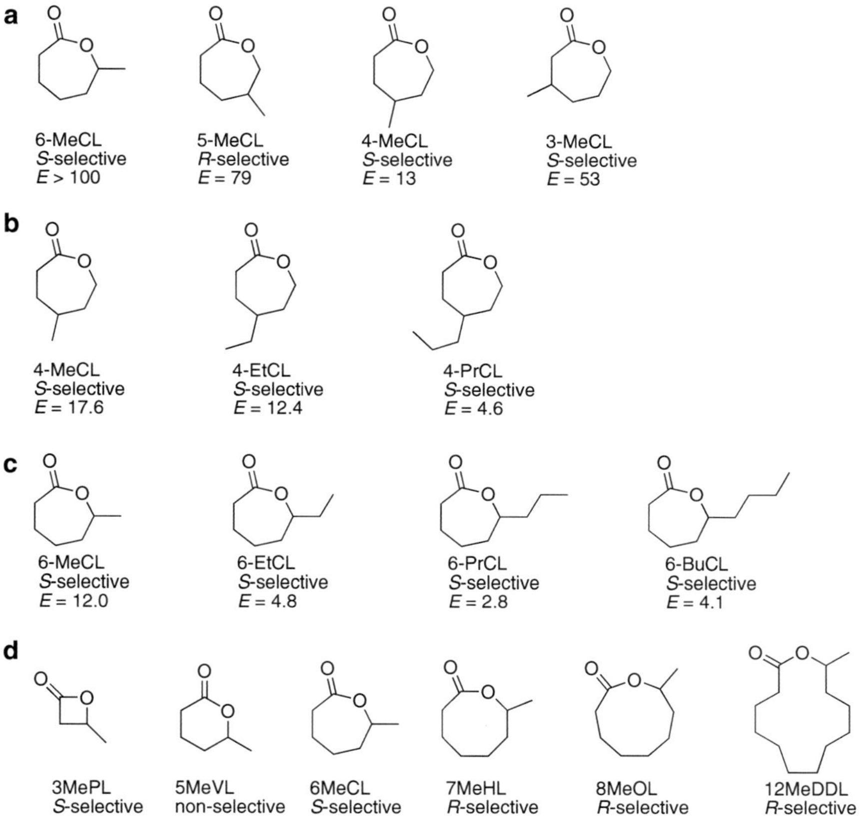

Fig. 10 Enantiopreference of CALB-catalyzed (**a**) hydrolysis of methyl-substituted ε-caprolactones [82], (**b**) hydrolysis of 4-alkyl-substituted ε-caprolactones [83], (**c**) transesterification of 6-alkyl-substituted ε-caprolactones [84], and (**d**) transesterification of ω-methyl-substituted lactones [85]

Shioji et al. found a pronounced (*S*)-selectivity in the hydrolysis of 6-methyl-ε-caprolactone (6-MeCL) employing CALB as the biocatalyst ($E > 100$) [82]. Hydrolysis of CLs with substituents at different positions (Fig. 10a) showed that there is an odd–even effect in the selectivity of CALB-catalyzed hydrolysis of methyl-substituted CLs, and *E* was moderate to good. Similar odd–even effects have been observed in the esterification of methyl-substituted decanoic acids when using *Candida rugosa* lipase [77]. Peeters et al. studied the hydrolysis of 4-alkyl substituted CLs (Fig. 10b) employing Novozym 435 [83]. Increasing the substituent size at the 4-position led to a rapid decrease of *E* from 17.6 to 4.6, although (*S*)-selectivity was observed in all cases.

Transesterification reactions of substituted lactones were also studied. The butanolysis of 6-MeCL evaluated by Bisht and coworkers in organic medium employing CALB showed an (*S*)-preference, but with a significantly lower selectivity ($E = 7$) compared to the hydrolysis reaction [86]. van As et al. studied the transesterification of 6-alkyl substituted CLs (Fig. 10c) [27, 84]. A decrease in the enantioselectivity was observed in the transesterification of 6-substituted CLs upon increasing the substituent size, but (*S*)-selectivity was present in all lactones. van Buitenen et al. evaluated the transesterification of lactones differing in ring size with the methyl substituent at the ω-position (Fig. 10d) [85]. Remarkably, small lactones (up to seven-membered ring) showed reactivity for the (*S*)-enantiomer or for both enantiomers, whereas in the case of larger rings the (*R*)-enantiomer was the faster reacting (Fig. 11). In fact, the (*R*)-selectivity for the transesterification of ω-methyl-substituted lactones is what one would expect on the basis of the (*R*)-selectivity found in the esterification of secondary alcohols. The observed high (*R*)-enantioselectivity in the hydrolysis or transesterification of these large ring lactones can be explained by the fact that the enantioselective step in the reaction sequence is merely the reverse of the acylation of a chiral secondary alcohol because the enantioselective step in the reaction sequence goes through the same transition state [87].

The (*S*)-selectivity for 6-MeCL in Novozym-435-catalyzed hydrolysis and transesterification reactions was unexpected. Recent molecular modeling studies by Veld et al. suggest that the *cisoid* ester conformation in lactones of ring size below eight could be responsible for the switch in enantioselectivity [88]. In contrast, the ester

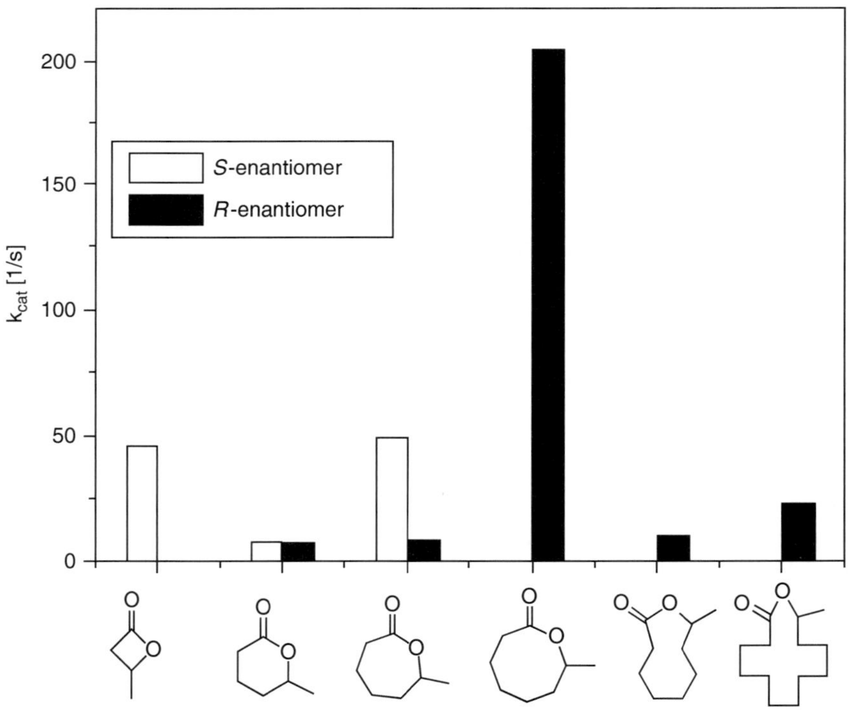

Fig. 11 Enantioselectivity in the CALB-catalyzed transesterification of ω-methyl lactones [75]

bond in larger ring lactones (ring size above eight) adopts a *transoid* conformation, as do open chain esters, explaining the observed (*R*)-selectivity [89]. This switch in selectivity has important implications in ROPs of ω-substituted lactones, as we will see in Sect. 3.3.

3.3 Enantioselective Ring-Opening Polymerizations of Substituted Lactones Using Lipases

Because the action of lipases relies on a two-step mechanism with an acylation step and a deacylation step, involving a covalent acyl-enzyme intermediate (Scheme 7), both the acylation and deacylation step can occur enantioselectively when using a (chiral) substituted lactone. In the case of a lactone as acyl donor, the product formed after ring-opening can act as a nucleophile, explaining the propensity of lactones to be polymerized by Novozym 435. Enantiodiscrimination occurs in the formation of the acyl-enzyme intermediate, but might also occur in the deacylation step because the alcohol formed after ring-opening, which serves as the propagating chain end, is chiral as well. It is well known that lipases show a pronounced selectivity for (*R*)-secondary alcohols in the deacylation step, but there are few reports dealing with the selectivity of substituents more remote from the nucleophilic center. Therefore, the enzymatic ROP of substituted lactones might result in optically active polymers, and the enantioselectivity observed could arise from the acylation step, the deacylation, or a combination of both.

The first enzymatic polymerizations of substituted lactones were performed by Kobayashi and coworkers using *Pseudomonas fluorescens* lipase or CALB as the biocatalyst [90–92]. A clear enantiopreference was observed for different lactone monomers, resulting in the formation of optically active polymers. More recently, a systematic study was performed by Al-Azemi et al. [93] and Peeters et al. [83] on the ROP of 4-alkyl-substituted CLs using Novozym 435. Peeters et al. studied the selectivity and the rates as a function of the substituent size with the aim of elucidating the mechanism and the rate-determining step in these polymerizations. Enantio-enriched polymers were obtained, but the selectivity decreased drastically with the increase in substituent size [83]. Remarkably for 4-propyl-ε-caprolactone, the selectivity was for the (*R*)-enantiomer in a polymerization, whereas it was (*S*)-selective in the hydrolysis reaction. Comparison of the selectivity in the hydrolysis reaction (Fig. 10b) with that of the polymerization reaction (Scheme 8a) revealed that the more bulky the alkyl substituent, the more important the deacylation step becomes as the rate-determining step.

The polymerization of lactones with the methyl substituent at different positions showed identical enantiopreferences in the polymerization reactions (Scheme 8b) as observed for the hydrolysis reaction (Fig. 10a) [26]. Also in this case, the alternating orientation of the methyl group from 3- to 5-MeCL in the faster reacting enantiomer suggested an odd–even effect. Comparison of the initial rate constants showed that the polymerization of 5-MeCL proceeded the fastest ($k_i = 5\,h^{-1}$), whereas 3-MeCL

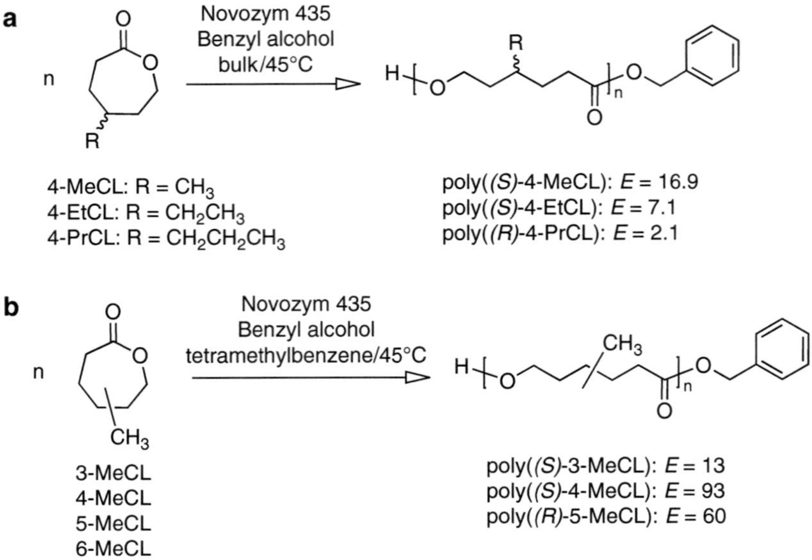

Scheme 8 Enzymatic polymerization of various substituted ε-caprolactones

and 4-MeCL showed moderate polymerization kinetics with initial rate constants of 1.3 and 1.1 h^{-1}, respectively. On the other hand, 6-MeCL did not polymerize ($k_i < 0.1\,h^{-1}$). Although the initiation (i.e., ring opening of 6-MeCL with the initiator benzyl alcohol) clearly occurred, the absence of further consumption of 6-MeCL was indicative of the virtual absence of propagation. This was explained by the configuration of the secondary alcohol formed after the ring opening of (*S*)-6-MeCL. The formed (*S*)-alcohol is unreactive in CALB-catalyzed esterifications of secondary alcohols (see above), hampering polymerization of this monomer on a practical time scale.

van Buijtenen et al. investigated the ROP of lactones with different ring sizes. In contrast to the ring-opening of the lactones with ring sizes of seven or less, ring-opening of the larger lactones with ring sizes between 8 and 13 was (*R*)-selective [85]. The formation of a reactive (*R*)-secondary alcohol for ω-methylated lactones with ring sizes of eight or more enabled straightforward kinetic resolution polymerization of these substrates. 7-Methylhepanolactone (7-MeHL), 8-methyloctanolactone (8-MeOL) and 12-methyldodecanolactone (12-MeDDL) (Fig. 10d) were successfully polymerized. The highest activity was observed for the polymerization of 7-MeHL ($k_{cat} = 270\,s^{-1}$), followed closely by that of 12-MeDDL ($k_{cat} = 223\,s^{-1}$). The polymerization of 8-MeOL was around five times slower ($k_{cat} = 44\,s^{-1}$) than the polymerization of 7-MeHL and 12-MeDDL. In all cases, good molecular weights were obtained for the chiral polymers (14–16 kDa), with polydispersity indexes (PDIs) varying between 1.23 and 2.25.

3.4 Chiral Polymers by Chemoenzymatic Catalysis

3.4.1 Dynamic Kinetic Resolution and its Extension to Polymer Chemistry

A prominent example of chemoenzymatic catalysis in bio-organic chemistry is the dynamic kinetic resolution (DKR) of secondary alcohols (Scheme 9) [94, 95] and amines [96–99]. In this process, a lipase is employed as an enantioselective acylation catalyst, and a metal-based catalyst ensures continuous racemization of the unreactive enantiomer.

Extension of DKR to polymer chemistry would readily result in chiral polyesters, polycarbonates, or polyamides from an optically inactive monomer mixture. Scheme 10 describes three variants of chemoenzymatic catalysis applied in polymer chemistry that recently appeared in the literature. Route A uses AA and BB monomers to prepare chiral polymers from racemic/diasteromeric diols. Route B converts an enantiomer mixture of AB monomers to homochiral polymers. Route C is the enzymatic ring-opening polymerization of ω-methylated lactones to homochiral polyesters. Details will be given in Sect. 3.4.2.

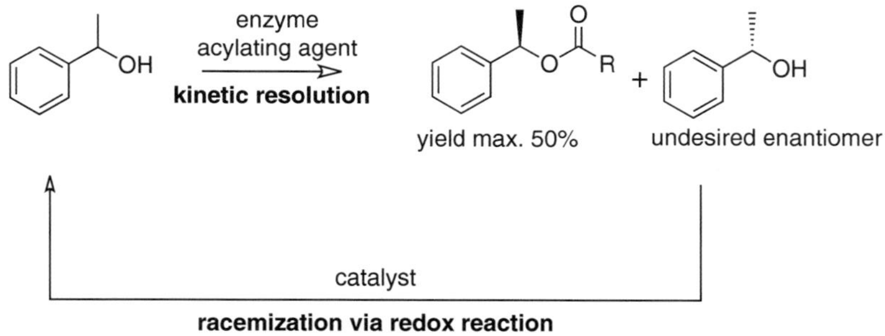

Scheme 9 Chemoenzymatic DKR of secondary alcohols

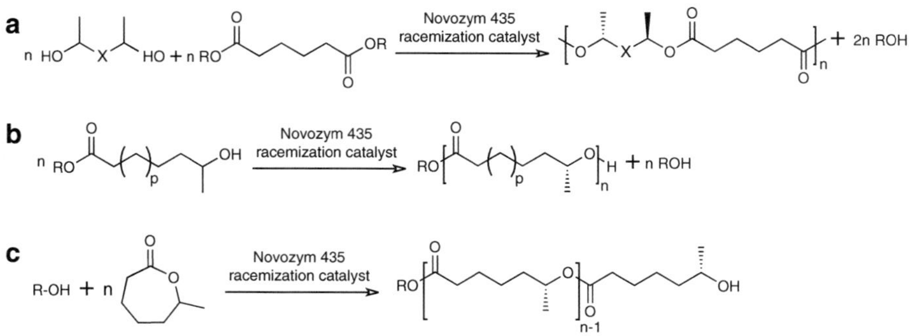

Scheme 10 Three variants (**a** – **c**) of chemoenzymatic DKR applied in polymer chemistry

3.4.2 Dynamic Kinetic Resolution Polymerization of AA–BB Monomers and AB Monomers

A straightforward extension of DKR to polymer chemistry is the use of diols and diesters (AA–BB monomers) or ester–alcohols (AB monomers) as substrates (Scheme 10, routes A and B). Such reactions have been referred to as DKR polymerizations and lead to the formation of oligomers and/or polymers because of the bifunctional nature of the reagents.

The extension of DKR to polymer chemistry is not trivial in practice since side reactions that are relatively unimportant in DKR (dehydrogenation, hydrolysis) have a major impact on the rate of polymerization and attainable chain lengths because the stoichiometry of the reactants is an important issue. As a result, the reaction conditions and catalyst combinations used in a typical DKR process will not a priori lead to chiral polymers from racemic or achiral monomers with good molecular weight (>10 kDa) and high *ee* (>95%).

Hilker et al. successfully employed the DKR of secondary alcohols developed at the Dutch company DSM [100] to prepare chiral polymers from α, α′-dimethyl-1,4-benzenedimethanol (1,4-diol) and dimethyl adipate (DMA) (Fig. 12a) [101].

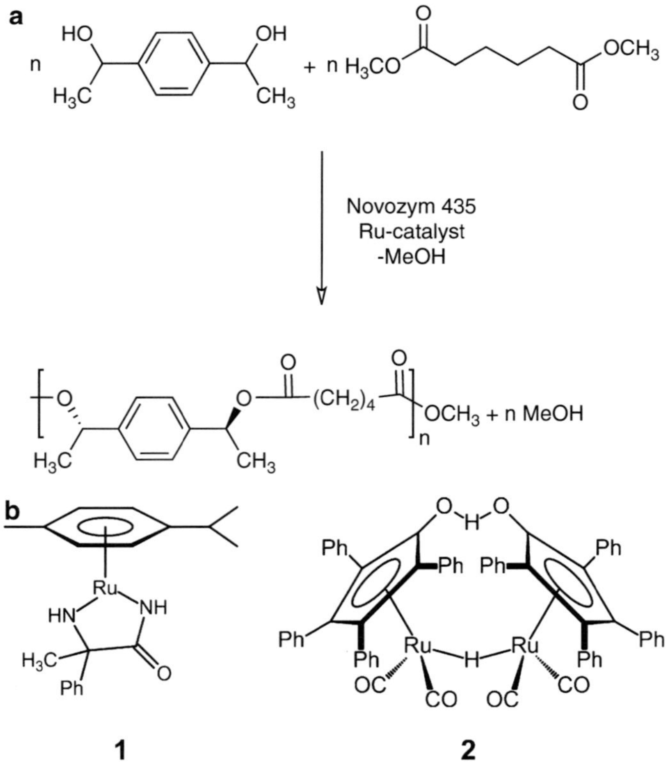

Fig. 12 (**a**) DKR polymerization of 1,4-diol and dimethyl adipate. (**b**) Chemical structure of Noyori-type racemization catalyst **1** and Shvo's racemization catalyst **2**

The applied catalytic system consisted of a Ru-Noyori-type racemization catalyst **1** (Fig. 12b) and Novozym 435. This catalyst combination tolerates a wide range of acyl donors, and it was expected that it would allow the use of bifunctional acyl donors for the formation of polycondensates. Before the start of the reaction, the monomer mixture showed the expected diastereomer ratio of $(S,S):(R,R):(R,S)$ of 1:1:2 of the 1,4-diol employed. After 30 h of reaction the (S,S)-enantiomer almost completely disappeared, whereas the ratio of (R,R)- to (R,S)-monomer was ca. 3:1 ($R:S$ ca. 7:1). At a hydroxyl group conversion of 92% after 70 h, no further conversion was observed and a final ratio of (R,R) to (R,S) of 16:1 ($R:S$ ca. 33:1) was obtained. Unfortunately, the molecular weights of the polymer were moderate at best ($M_w = 3.4$ kDa) and Novozym 435 had to be added every few hours to compensate for the activity loss of the lipase. This suggests that Ru-catalyst **1** and Novozym 435 are not fully compatible.

van As et al. improved the catalytic system by changing the racemization catalyst from **1** to Shvo's catalyst **2** [102] (Fig. 12b) and by adding the hydrogen donor 2,4-dimethyl-3-pentanol (DMP) to suppress dehydrogenation reactions [103]. DMP is a sterically hindered alcohol and is not accepted by Novozym 435 as a substrate. Although Shvo's catalyst **2** is a significantly slower racemization catalyst than **1**, it does not require the addition of K_2CO_3. This base, required to activate the precatalyst of **1**, appears to contribute to the deactivation of Novozym 435 in DKR polymerization conditions. For the 1,4-diol, the improved catalyst combination resulted in polymers with *ee* of 94% and peak molecular weight (M_p) of 8.3 kDa within 170 h. α,α'-Dimethyl-1,3-benzenedimethanol (1,3-diol), showed a higher solubility in toluene, and slightly better results under similar conditions were obtained. After an optimization study using 1,3-diol and diisopropyl adipate as the monomers, chiral polymers were obtained with M_p values up to 15 kDa, *ee* values up to 99%, and (at most) 1–3% ketone functional groups in ~120 h as a result of dehydrogenation. Aliphatic secondary diols were also employed as the substrate, but DKR polymerization of these diols did not lead to enantiopure polymers. At most, an *ee* of 46% was obtained with low molecular weights in the range of 3.3–3.7 kDa. The latter was attributed to the low selectivity of Novozym 435 for these secondary diols, as revealed by kinetic resolution experiments of 2,9-decandiol with vinyl acetate and Novozym 435. Apparently, the (S)-alcohol showed significant reactivity, decreasing the *ee* of the polymer.

In contrast, Kanca et al. observed that aliphatic AB monomers (Scheme 11) comprising a secondary hydroxyl group and an ester moiety did show high enantioselectivities in a Novozym-435-catalyzed transesterification reaction [104]. The E was high for all monomers studied ($E > 200$). An additional advantage

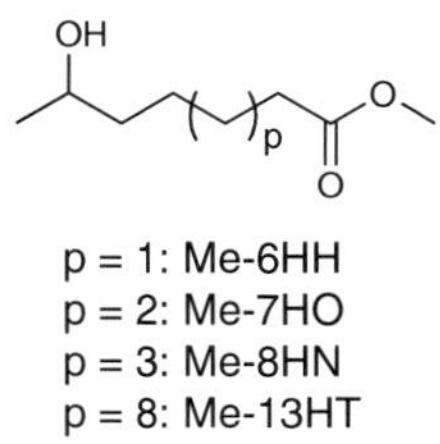

Scheme 11 Novozym-435-catalyzed transesterification reaction of aliphatic AB monomers containing a secondary hydroxyl group [104]

of the use of AB monomers over AA–BB monomers is the lack of sensitivity to stoichiometric issues. Polymerization of the monomers (Scheme 10, route B) employing Novozym 435 and Ru-catalyst **2** was slow (170 h) but did result in chiral polymers of good molecular weight and *ee*. For example, polymerization of Me-7HO resulted in a polymer with an M_p of 16.3 kDa and an *ee* of 92%.

3.4.3 Chiral Polymers from Racemic ω-Methylated Lactones

As described above, 6-MeCL was found to be unreactive in a Novozym-435-catalyzed polymerization reaction as a result of the formation of an unreactive (*S*)-alcohol chain. These (*S*)-alcohols are not accepted as a nucleophile by Novozym 435 since the lipase-catalyzed transesterification is highly (*R*)-selective for the nucleophile. van As et al. proposed that by combining Novozym-435-catalyzed ring-opening of 6-MeCL with racemization of the terminal alcohol, it should be possible to polymerize 6-MeCL [105]. The method was termed iterative tandem catalysis (ITC) since chain propagation can only be achieved by a combination of two fundamentally different catalytic systems. The proof-of-principle was shown by the polymerization of (*S*)-6-MeCL using Novozym 435 and Noyori-type catalyst **1** (Scheme 12). The ring-opening of (*S*)-6-MeCL, the preferred enantiomer

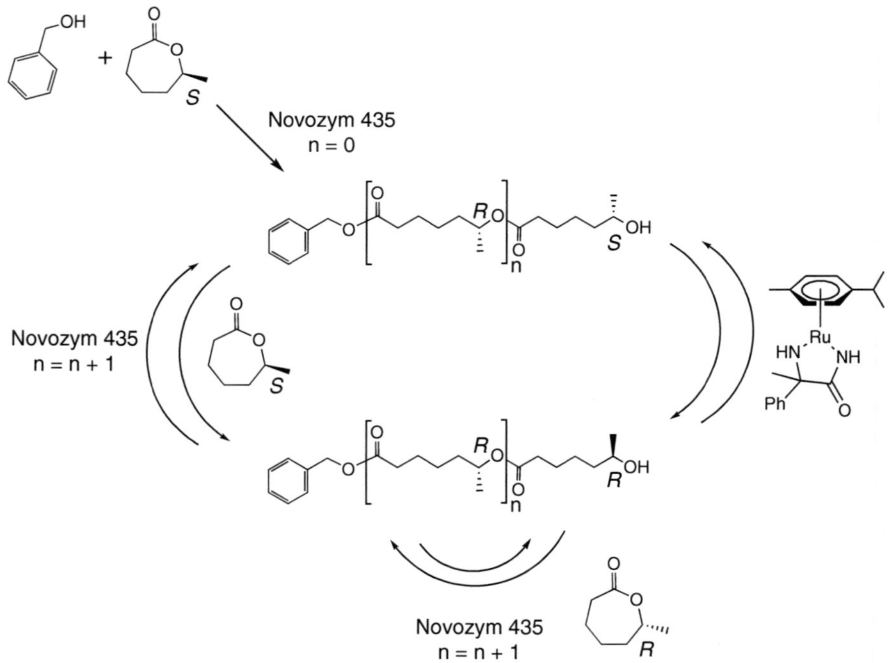

Scheme 12 Chiral polymers from racemic ω-methylated lactones by iterative tandem catalysis [105, 106]

of Novozym 435, will lead to a ring-opened product with an (*S*)-alcohol chain end. This product is virtually unreactive and propagation does not occur. To enable polymerization of (*S*)-6-MeCL, racemization of the (*S*)-alcohol that is formed upon ring-opening is required, furnishing a reactive (*R*)-chain end. These can subsequently react with another molecule of (*S*)-6-MeCL. If the less reactive (*R*)-6-MeCL is incorporated (which will occur because the selectivity for lactones is moderate with an *E* of 12), a reactive (*R*)-chain end is obtained and propagation occurs instantly. In this way, both enantiomers of the monomer are consumed. It appeared, however, that polymerization with the Novozym 435/**1** catalytic system was not feasible: only oligomers were obtained in one-step reactions because the catalysts were poorly compatible under the reaction conditions employed.

A significant improvement was achieved when Shvo's Ru-catalyst **2** (Fig. 12b) was employed in combination with the addition of DMP to suppress dehydrogenation reactions [106]. Poly-(*R*)-6-MeCL, with a promising *ee* of 86% and M_p of 8.2 kDa, was obtained after workup starting from optically pure (*S*)-6-MeCL. The low rate of reaction compared to DKR (typically complete after 48 h with the Shvo catalyst) is attributed to the low concentration of the terminal alcohol as well as to the iterative nature of the system. Racemic 6-MeCL showed comparable rates of reaction for both enantiomers, which polymerized within 220 h with complete conversion of both enantiomers, yielding polymers of high *ee* (92%) and M_p (9.4 kDa). Successful polymerizations with more than 100 consecutive and iterative enzymatic additions and Ru-catalyzed racemizations on one polymer chain were realized.

3.4.4 Chiral Block Copolymers by Chemoenzymatic Catalysis

Chiral polymers can be prepared using a one-pot system, i.e., all reactants and catalysts are present at the start of the reaction and both catalysts work simultaneously. However, one can also envisage the synthesis of chiral polymers using catalysts in sequence, either in one pot or even completely independent of each other. This section will deal with the synthesis of chiral block copolymers using different catalysts in sequence. An interesting example of the synthesis of chiral polymers using catalysts in sequence is the synthesis of chiral block copolymers in a sequential approach. Both ATRP and nitroxide-mediated LFRP were evaluated for this purpose.

Peeters et al. combined the enzymatic ring-opening polymerization of 4-MeCL with a controlled ATRP of MMA (Scheme 13) [27]. This resulted in the formation of a chiral block copolymer. In the case of combining ATRP and enantioselective ROP (eROP) of 4-MeCL, it was found that the addition of $Ni(PPh_3)_4$ inhibited Novozym 435 and at the same time catalyzed the ATRP reaction. While Novozym 435 did not interfere with the ATRP of MMA, $Ni(PPh_3)_4$ clearly inhibited Novozym 435 during the eROP of 4-MeCL. After precipitation, the chiral block copolymers ($M_n = 11$ kDa and 17 kDa) were isolated as solid compounds showing two glass transition temperatures (T_g) at -60 and 100 °C, indicative of phase separation between the two blocks.

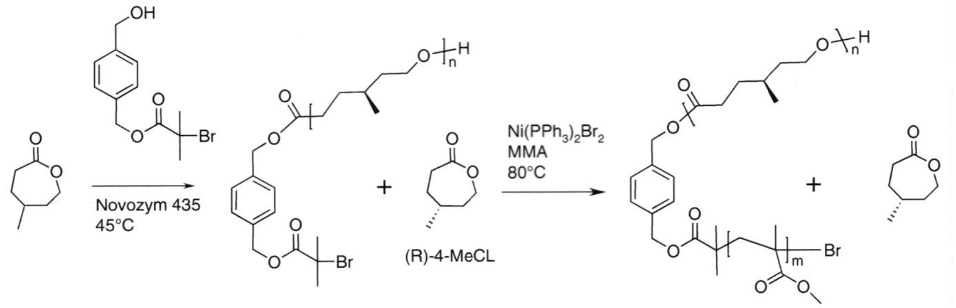

Scheme 13 Synthesis of a chiral block copolymer by combining ATRP and enantioselective enzymatic ROP of 4-MeCL [27]

A similar approach was followed with the eROP of 4-MeCL, followed by nitroxide-mediated LFRP of styrene using a bifunctional catalyst (Scheme 4) [43]. Styrene, the monomer for the NMP was added at the beginning because it proved to be a good solvent for an eROP. An inhibitor, paraoxon, was added after the eROP to prevent further transesterification reactions, and increasing the temperature to 95 °C started the nitroxide-mediated LFRP. After precipitation, the chiral block copolymers obtained showed two T_g at −51 °C and 106 °C. The specific rotation $[\alpha]_D^{25}$ of the block copolymer was −2.6°. This is in good agreement with the optical rotation of −7.2° reported by Bisht [93] for poly[(*S*)-4-MeCL] with an *ee* of 90%, considering that the block length ratio of chiral to nonchiral block in the poly[(*S*)-4-MeCL-*co*-S] block copolymer is ca. 1:2.

3.4.5 Chiral Side-Chain Polymers by Chemoenzymatic Catalysis

Duxbury et al. recently showed an elegant approach for the tuning of polymer properties with chirality [107]. With the aid of two alcohol dehydrogenases that show opposite enantioselectivities in the reduction of ketones (ADH-LB and ADH-T), the two enantiomers of *p*-vinylphenylethanol were obtained in excellent yield and *ee*. Copolymers of these monomers with styrene using free-radical polymerization afforded random block copolymers with compositions ranging from 100% (*R*)- to 100% (*S*)-alcohol in the side chains (Scheme 14). The backbones were designed to contain a total of approximately 45% of the secondary alcohol monomer (either (*R*), (*S*), or a mixture). The polymers showed number-average molecular weight M_n of 5.0–6.0 kDa and PDIs between 1.7 and 2.1. From differential scanning calorimetry (DSC) analysis, a similar T_g of approximately 115 °C was found for all polymers, irrespective of their chiral composition, suggesting that the enantiomeric composition has no effect on their thermal properties. The optical rotation of the polymers increased linearly from −20 to +20°. Novozym 435 was used to catalyze the esterification of the alcohol groups on the polymer backbone with vinyl acetate in toluene (Scheme 14). When a backbone containing 100% (*S*)-secondary alcohol groups (styrene copolymer containing approximately 45% alcohol monomer) was

Scheme 14 Tuning of polymer properties by enantioselective enzymatic transesterification of chiral polymers [107]

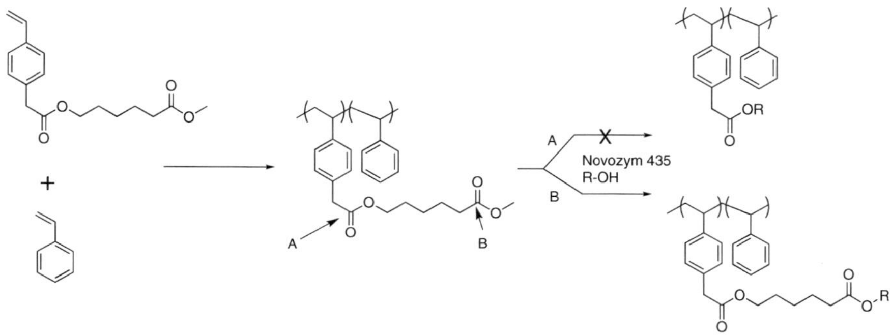

Scheme 15 Regioselective transesterification of polymer side-group esters [108]

used for the enzymatic grafting of vinyl acetate, no reaction was observed over a period of 24 h. In contrast, when a backbone containing 100% (*R*)-secondary alcohol groups was used, the enzymatic esterification of vinyl acetate occurred from 75% of the alcohol groups within 24 h. This is the first example in which chiral information stored in a polymer chain can be "read out" by an enantioselective enzymatic reaction.

In a related approach, Padovani et al. prepared copolymers of styrene and a styrene derivative containing two pendant ester bonds using free-radical polymerization (Scheme 15) [108]. Transesterification reactions were conducted with Novozym 435 as the catalyst and benzyl alcohol or (*rac*)-1-phenylethanol as the nucleophile. Interestingly, the ester bond closest to the polymer backbone (position A in Scheme 15) remained unaffected, whereas ester bond B reacted in up to 98% to the corresponding benzyl ester. The transesterification was not only highly chemoselective but also enantioselective. Conversion of (*rac*)-1-phenylethanol in the transesterification reaction amounted to a maximum conversion of 47.9% of the (*R*)-alcohol, and only at the ester position B.

Yashima et al. showed an example where the polymer helicity was controlled by enzymatic enantioselective acylation of the monomers [109]. Optically active phenylacetylenes containing hydroxyl or ester groups were obtained by the kinetic resolution of the corresponding racemic hydroxy-functional phenylacetylene (see Scheme 16). Polymerization of the phenylacetylenes afforded an optically active poly(phenylacetylene) with a high molecular weight ($M_n = 89$ kDa; PDI = 2.0) and

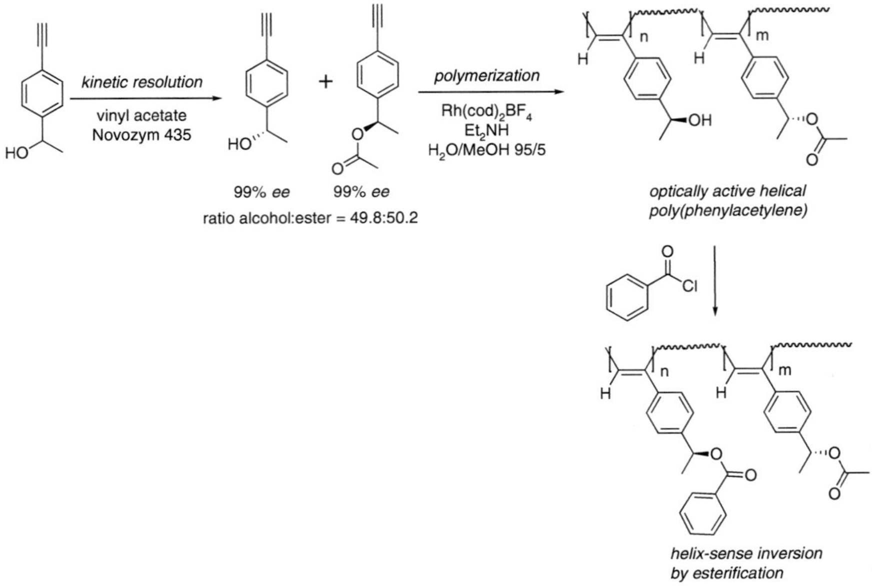

Scheme 16 Synthesis of optically active poly(phenylacetylene) from optically active phenylacetylenes [109]

with a preferred helical sense, as evidenced by circular dichroism spectroscopy. Poly(acetylenes) are dynamic helical polymers composed of interconvertible left- and right-handed helical conformations separated by helical reversals. The optical activity introduced by the small excess of chiral monomers biases one of the helical conformations. Upon reaction of the chiral alcohols in the polymer using an acid chloride or an isocyanate, the helix sense was inverted. Although helix inversions have been reported previously for poly(phenylacetylenes) by temperature, solvent, and chiral additives, this is the first example of helicity inversion by chemical modification using achiral compounds.

4 Conclusions

Enzymatic polymerizations have been established as a promising and versatile technique in the synthetic toolbox of polymer chemists. The applicability of this technique for homo- and copolymerizations has been known for some time. With the increasing number of reports on the synthesis of more complex structures like block copolymers, graft copolymers, chiral (co)polymers, and chiral crosslinked nanoparticles, its potential further increases. Although not a controlled polymerization technique itself, clever reaction design and integration with other polymerization techniques like controlled radical polymerization allows the procurement of well-defined polymer structures. Specific unique attributes of the enzyme can be applied to develop materials and processes difficult to achieve otherwise. Examples are the synthesis of chiral block copolymers (stereoselectivity), the selective modification of polymer endgroups using enzymatic chemoselectivity, the synthesis of fluorinated block copolymers in a fully green process in $scCO_2$, and the synthesis of optically active materials from racemic mixtures in one step. It can be expected that this trend will continue and we will see more examples in which the specific strength of the enzyme (selectivity) leads to unique materials. As a result, a combined multidisciplinary effort at the interface of biocatalysis, polymer chemistry, and organic catalysis could lead to a paradigm shift in polymer chemistry and allow a higher level of structural complexity in macromolecules, reminiscent of those found in Nature.

References

1. Bruggink A, Schoevaart R, Kieboom T (2003) Org Process Res Dev 7:622
2. Gross RA, Kumar A, Kalra B (2001) Chem Rev 101:2097
3. Kobayashi S (2009) Macromol Rapid Commun 30:237
4. Bernaerts KV, Du Prez FE (2006) Prog Polym Sci 31:671
5. Xiao Y, Cummins D, Palmans ARA, Koning CE, Heise A (2008) Soft Matter 4:593
6. Takwa M, Xiao Y, Simpson N, Malmstrom E, Hult K, Koning CE, Heise A, Martinelle M (2008) Biomacromolecules 9:704
7. Xiao Y, Takwa M, Hult K, Koning CE, Heise A, Martinelle M (2009) Macromol Biosci 9:713
8. Takwa M, Hult K, Martinelle M (2008) Macromolecules 41:5230
9. Simpson N, Takwa M, Hult K, Johansson M, Martinelle M, Malmstrom E (2008) Macromolecules 41:3613
10. Habaue S, Asai M, Morita M, Okamoto Y, Uyama H, Kobayashi S (2003) Polymer 44:5195
11. van der Meulen I, de Geus M, Antheunis H, Deumens R, Joosten EAJ, Koning CE, Heise A (2008) Biomacromolecules 9:3404
12. Uyama H, Kobayashi S, Morita M, Habaue S, Okamoto Y (2001) Macromolecules 34:6554
13. Duda A, Kowalski A, Penczek S, Uyama H, Kobayashi S (2002) Macromolecules 35:4266
14. Kobayashi S, Sakamoto J, Kimura S (2001) Prog Polym Sci 26:1525
15. Popescu D, Keul H, Moeller M (2009) Macromol Chem Phys 210:123
16. Kumar A, Gross RA, Wang YB, Hillmyer MA (2002) Macromolecules 35:7606
17. Panova AA, Kaplan DL (2003) Biotechnol Bioeng 84:103
18. He F, Li SM, Vert M, Zhuo RX (2003) Polymer 44:5145
19. Srivastava RK, Albertsson AC (2006) Macromolecules 39:46
20. Kaihara S, Fisher JP, Matsumura S (2009) Macromol Biosci 9:613
21. Sharma B, Azim A, Azim H, Gross RA, Zini E, Focarete L, Scandola M (2007) Macromolecules 40:7919
22. Niu L, Nagarajan R, Guan F, Samuelson LA, Kumar J (2006) J Macromol Sci Pure Appl Chem 43:1975
23. Meyer U, Palmans ARA, Loontjens T, Heise A (2002) Macromolecules 35:2873
24. de Geus M, Peeters J, Wolffs M, Hermans T, Palmans ARA, Koning CE, Heise A (2005) Macromolecules 38:4220
25. de Geus M, Peters R, Koning CE, Heise A (2008) Biomacromolecules 9:752
26. De Geus M, Schormans L, Palmans AA, Koning CE, Heise A (2006) J Polym Sci A Polym Chem 44:4290
27. Peeters J, Palmans ARA, Veld M, Scheijen F, Heise A, Meijer EW (2004) Biomacromolecules 5:1862

28. Duxbury CJ, Wang WX, de Geus M, Heise A, Howdle SM (2005) J Am Chem Soc 127:2384
29. Zhou J, Villarroya S, Wang W, Wyatt MF, Duxbury CJ, Thurecht KJ, Howdle SM (2006) Macromolecules 39:5352
30. Villarroya S, Thurecht KJ, Heise A, Howdle SM (2007) Chem Commun 3805
31. Sha K, Li DS, Wang SW, Qin L, Wang JY (2005) Polym Bull 55:349
32. Sha K, Li D, Li Y, Ai P, Wang W, Xu Y, Liu X, Wu M, Wang S, Zhang B, Wang J (2006) Polymer 47:4292
33. Sha K, Li D, Li Y, Liu X, Wang S, Wang J (2008) Polym Int 57:211
34. Sha K, Li DS, Li YP, Ai P, Liu XT, Wang W, Xu YX, Wang SW, Wu MZ, Zhang B, Wang JY (2006) J Polym Sci A Polym Chem 44:3393
35. Li M, Wang W, Pan S, Zhang B, Wang J (2008) Polym Int 57:1377
36. Sha K, Qin L, Li DS, Liu XT, Wang JY (2005) Polym Bull 54:1
37. Sha K, Li D, Li Y, Liu X, Wang S, Guan J, Wang J (2007) J Polym Sci A Polym Chem 45:5037
38. Li D, Sha K, Li Y, Ai P, Liu X, Wang W, Wang J (2008) Polym Int 57:571
39. Sha K, Li D, Li Y, Zhang B, Wang J (2008) Macromolecules 41:361
40. Zhang B, Li Y, Xu Y, Wang S, Ma L, Wang J (2008) Polym Bull 60:733
41. Zhang B, Li Y, Wang W, Chen L, Wang S, Wang J (2009) Polym Bull 62:643
42. Zhang B, Li Y, Sun J, Wang S, Zhao Y, Wu Z (2009) Polym Int 58:752
43. van As BAC, Thomassen P, Kalra B, Gross RA, Meijer EW, Palmans ARA, Heise A (2004) Macromolecules 37:8973
44. Thurecht KJ, Gregory AM, Villarroya S, Zhou J, Heise A, Howdle SM (2006) Chem Commun 4383
45. Kerep P, Ritter H (2007) Macromol Rapid Commun 28:759
46. Deng F, Bisht KS, Gross RA, Kaplan DL (1999) Macromolecules 32:5159
47. Xiao Y, Coulembier O, Koning CE, Heise A, Dubois P (2009) Chem Commun 2472
48. Xue L, Dai S, Li Z (2009) Macromolecules 42:964
49. Kalra B, Kumar A, Gross RA, Baiardo M, Scandola M (2004) Macromolecules 37:1243
50. Srivastava RK, Kumar K, Varma IK, Albertsson A (2007) Eur Polym J 43:808
51. Keul H, Moeller M (2009) J Polym Sci A Polym Chem 47:3209
52. Hans M, Gasteier P, Keul H, Moeller M (2006) Macromolecules 39:3184
53. Duxbury CJ, Cummins D, Heise A (2007) Macromol Rapid Commun 28:235
54. Hans M, Keul H, Heise A, Moeller M (2007) Macromolecules 40:8872
55. Hans M, Xiao Y, Keul H, Heise A, Moeller M (2009) Macromol Chem Phys 210:736
56. Villarroya S, Zhou J, Thurecht KJ, Howdle SM (2006) Macromolecules 39:9080
57. Villarroya S, Dudek K, Zhou J, Irvine DJ, Howdle SM (2008) J Mater Chem 18:989
58. Peeters JW, Palmans ARA, Meijer EW, Koning CE, Heise A (2005) Macromol Rapid Commun 26:684
59. Okamoto Y, Nakano T (1994) Chem Rev 94:349
60. Dechy-Cabaret O, Martin-Vanca B, Bourissou D (2004) Chem Rev 104:6147
61. Spassky N, Wisniewsky M, Pluta Ch, Le Borgne A (1996) Macromol Chem Phys 197:2627
62. Amgoune A, Thomas CM, Ilinca S, Roisnel T, Carpentier J-F (2006) Angew Chem Int Ed 45:2782
63. Ovitt TM, Coates GW (1999) J Am Chem Soc 121:4072
64. Zhong Z, Dijkstra PJ, Feijen J (2003) J Am Chem Soc 125:11291
65. Chen C-S, Wu S-H, Girdaukas G, Sih CJ (1987) J Am Chem Soc 109:2812
66. Faber K (2004) Biotransformation in organic chemistry, 5th edn. Springer, Berlin, p 42
67. Nardini M, Dijkstra BE (1999) Curr Opin Struct Biol 9:732
68. Orrenius C, Haeffner F, Rotticci D, Ohrner N, Norin T, Hult K (1998) Biocatal Biotransform 16:1
69. Patel RN, Banerjee A, Nanduri V, Goswami A, Comezoglu FT (2000) J Am Oil Chem Soc 77:1015
70. Ottosson J, Hult K (2001) J Mol Catal B Enzyme 11:1025
71. Rotticci D, Haeffner F, Orrenius C, Norin T, Hult K (1998) J Mol Catal B Enzyme 5:267
72. Magnusson AO, Takwa M, Hamburg A, Hult K (2005) Angew Chem Int Ed 44:2

73. Kazlauskas R, Weissfloch ANE, Rappaport AT, Cuccia LA (1991) J Org Chem 56:2656
74. Uppenberg J, Ohrner N, Norin M, Hult K, Kleywegt GJ, Patkar S, Waagen V, Anthonsen T, Jones TA (1995) Biochemistry 34:16838
75. Arroyo M, Sinisterra J (1994) J Org Chem 59:4410
76. Pepin P, Lortie R (1999) Biotechnol Bioeng 63:502
77. Hedenstrom E, Nguyen B-V, Silks III LA (2002) Tetrahedron Asymmetry 13:835
78. Reetz MT, Wilensek S, Zha D, Jaeger K-E (2001) Angew Chem Int Ed 40:3589
79. Holmquist M, Berglund P (1999) Org Lett 1:763
80. Bocola M, Otte N, Jaeger K-E, Reetz MT, Thiel W (2004) ChemBioChem 5:214
81. Huisgen R, Ott H (1959) Tetrahedron 6:253
82. Shioji K, Matsuo A, Okuma K, Nakamura K, Ohno A (2000) Tetrahedron Lett 41:8799
83. Peeters JW, van Leeuwen O, Palmans ARA, Meijer EW (2005) Macromolecules 38:5587
84. Palmans ARA, van As BAC, van Buijtenen J, Meijer EW (2008) In: Cheng HN, Gross RA (eds) Polymer biocatalysis and biomaterials II. ACS Symp Ser 999:230–244
85. van Buijtenen J, van As BAC, Verbruggen M, Roumen L, Vekemans JAJM, Pieterse K, Hilbers PAJ, Hulshof LA, Palmans ARA, Meijer EW (2007) J Am Chem Soc 129:7393
86. Kondaveti L, Al-Azemi T, Bisht KS (2002) Tetrahedron Asymmetry 13:129
87. Chen C, Fujimoto Y, Girdaukas G, Sih CJ (1982) J Am Chem Soc 104:7294
88. Veld MAJ, Fransson L, Meijer EW, Palmans ARA, Hult K (2009) ChemBioChem 10:1330
89. van As BAC, Chan D-K, Kivit PJJ, Palmans ARA, Meijer EW (2007) Tetrahedron Asymmetry 18:787
90. Kuellmer K, Kikuchi H, Uyama H, Kobayashi S (1998) Macromol Rapid Commun 19:127
91. Kikuchi H, Uyama H, Kobayashi S (2000) Macromolecules 33:8971
92. Kikuchi H, Uyama H, Kobayashi S (2002) Polym J 34:835
93. Al-Azemi TF, Kondaveti L, Bisht KS (2002) Macromolecules 35:3380
94. Pamiès O, Bäckvall J-E (2003) Chem Rev 103:3247
95. Pamiès O, Bäckvall J-E (2004) Trends Biotechnol 22:130
96. Reetz MT, Schimossek K (1996) Chimia 50:668
97. Paetzold J, Bäckvall J-E (2005) J Am Chem Soc 127:17620
98. Gastaldi S, Escoubet S, Vanthuyne N, Gil G, Bertrand MP (2007) Org Lett 9:837
99. Veld MAJ, Hult K, Palmans ARA, Meijer EW (2007) Eur J Org Chem 32:5416
100. Verzijl GKM, De Vries JG, Broxterman QB (2001) PCT Int Appl WO2001090396
101. Hilker I, Rabani G, Verzijl GKM, Palmans ARA, Heise A (2006) Angew Chem Int Ed 45:2130
102. Karvembu R, Phabharakan R, Natarajan K (2005) Coord Chem Rev 249:911
103. van As BAC, van Buijtenen J, Mes T, Palmans ARA, Meijer EW (2007) Chemistry 13:8325
104. Kanca U, van Buijtenen J, van As BAC, Korevaar P, Vekemans JAJM, Palmans ARA, Meijer EW (2008) J Pol Sci Part A Polym Chem 46:2721
105. van As BAC, van Buijtenen J, Heise A, Broxterman QB, Verzijl GKM, Palmans ARA, Meijer EW (2005) J Am Chem Soc 127:9964
106. van Buijtenen J, van As BAC, Meuldijk J, Palmans ARA, Vekemans JAJM, Hulshof LA, Meijer EW (2006) Chem Commun 30:3169
107. Duxbury CJ, Hilker I, de Wildeman SMA, Heise A (2007) Angew Chem Int Ed 46:8452
108. Padovani M, Hilker I, Duxbury CJ, Heise A (2008) Macromolecules 41:2439
109. Kobayashi S, Morino K, Yashima E (2007) Chem Commun 30:2351

Adv Polym Sci (2010) 237: 115–126
DOI:10.1007/12_2010_89

Published online: 26 June 2010

Hydrolases in Polymer Chemistry: Part III: Synthesis and Limited Surface Hydrolysis of Polyesters and Other Polymers

Georg M. Guebitz

Abstract Limited enzymatic surface hydrolysis of polyamides, polyethyleneterphthalates (PET) and polyacrylonitriles has been demonstrated to be a powerful and yet mild strategy for directly improving polymer surface properties (e.g., hydrophilicity) or activating materials for further processing. Recently, mechanistic details on enzymatic surface hydrolysis have become available, especially for the functionalisation of PET, which has been investigated in most detail. Generally, enzymes show a strong preference for amorphous regions of polymers. Consequently, during hydrolysis, the degree of crystallinity increases according to FTIR and DSC analysis. MALDI-TOF analysis has shown that PET hydrolases (i.e. cutinases and lipases) cleave the polymer *endo*-wise, in contrast to alkaline hydrolysis. As a result, an increase in the amount of carboxyl and hydroxyl groups has been found upon enzymatic hydrolysis, according to X-ray photoelectron spectroscopy and various derivatisation and titration methods recently adapted for this purpose. These mechanistic data, combined with advances in structural and molecular biology, help to explain the considerably different activities of closely related enzymes (e.g. cutinases) on polymers.

Keywords Cutinase · Polyacrylonitrile · Polyamide · Polyethyleneterphthalate · Surface hydrolysis

Contents

G.M. Guebitz (✉)
Department of Environmental Biotechnology, Graz University of Technology, Petersgasse 12, 8010 Graz, Austria
e-mail: guebitz@tugraz.at

1 Limited Surface Hydrolysis of Synthetic Polymers

1.1 General Aspects: Introduction

Limited enzymatic surface hydrolysis of polyamides (PA), polyethyleneterphthalates (PET) and polyacrylonitriles (PAN) has been demonstrated to be a powerful but mild strategy for directly improving polymer properties or for activating materials for further processing. In contrast to chemical hydrolysis, the enzyme-catalysed reaction does not change polymer bulk properties but specifically introduces novel functional groups (i.e. carboxyl, hydroxyl, amino, amide) on the surface [1] Thereby, a successful polyamidase or PET hydrolase catalyses a limited *endo*-wise hydrolysis without any release of oligomers or monomers [2]. Consequently, there will be no weight loss, which can reach up to 15% in chemical hydrolysis [1, 3, 4]. For PAN-hydrolyzing enzymes, only excessive hydrolysis should be avoided, which could potentially lead to the release of polyacrylic acid [5]. Generally, limited surface hydrolysis enhances the hydrophilicity of the polymers and materials made thereof. This is a desired characteristic for synthetic textile fabrics due to lower built up of static charges, and improved breathability, moisture transport and wearing comfort. In contrast to impregnation with chemical finishers for the same purpose, the covalent enzymatic modification obviously has a higher fastness. Improved hydrophilicity of PA can decrease the tendency of PA-based ultrafiltration membranes for fouling by proteins and other biomolecules. Thus, enzymatic hydrophilisation of such membranes could decrease the energy demand for filtration and for cleaning with aggressive chemicals or could avoid frequent replacement [6–9]. Similarly, increased hydrophilicity of PET (i.e. 15° lower contact angle) has been demonstrated to reduce bacterial adhesion and, consequently, reduce infections of cardiovascular implants such as artificial heart valve sewing rings and artificial blood vessels [9].

In addition to direct beneficial effects of increased hydrophilicity, this property is a key requirement in further processing of synthetic materials in gluing, painting, inking, anti-fogging, filtration, textiles, electronics and biomedical areas [10]. For example, an increased number of functional groups on the PET surface can enhance binding and reduce binder consumption in coating with PVC [11]. Similarly, in the production of flexible electronic devices (FEDs) such as displays or photovoltaic cells, surface hydrophilisation is required for the attachment of functional layers [12]. Many reports have shown that increased hydrophilicity of synthetic fabrics can improve dye binding and fastness. In Sect. 1.2, recent advances in enzymatic surface hydrolysis of PA, PET and PAN will be discussed with special focus on mechanistic aspects and substrate specificities of the various biocatalysts investigated for this purpose.

1.2 Monitoring Surface Hydrolysis

For various applications of synthetic polymers, such as in textiles, limited enzymatic surface hydrolysis should ultimately bring increases in hydrophilicity. Consequently, quantification of this parameter was the main target of early studies on enzymatic surface hydrolysis. Thereby, simple and fast analysis methods primarily involved measurement of the water contact angle (WCA) and the rising height, and the drop dissipation test. In addition, dye binding on textile fabrics was used as a measure to quantify success of the enzyme treatment (Table 1). However, besides enzymatic hydrolysis, enzyme protein adsorption can also increase the hydrophilicity and, consequently, the dye binding on synthetic polymers. For example, treatment of PET with a cutinase from *Fusarium solani* and a lipase from *Candida antarctica* resulted in protein layers with thickness of 1.6–2.6 nm and 2.5–2.8 nm, respectively, according to angle-resolved X-ray photoelectron spectroscopy (XPS) [13]. Therefore, either complete protein removal (e.g. via extraction, detergents or proteases) [1, 14] or the use of enzyme inhibitors in control experiments is an experimental prerequisite when the above analysis methods are used [15]. XPS was also used for monitoring PAN hydrolysis based on increasing

Table 1 Monitoring enzymatic surface hydrolysis of polymers

Method	Target	Polymers	References
FTIR	Crystallinity	PET, PA	[14, 19]
	Amino and amide groups		[16]
DSC	Crystallinity	PET, PTT	[15, 18]
XPS (ESCA)	Atomic composition	PET, PTT	[1, 2, 13, 16, 17]
	Carboxyl and hydroxyl groups, nitrogen content	PAN	
MALDI-TOF	Changes in DP	PET	[2]
	Oligomers formed		
SEM	Morphology	PET, PA, PAN	Various
WCA, drop test, rising height	Hydrophilicity	PET, PA, PAN	Various
Tensiometry	Surface charge	PET, PA	[11, 20]
Dye binding	Indirect measure for functional groups	PET, PA, PAN	Various
Derivatisation	Quantification of carboxyl and hydroxyl groups with 2-(bromoniethyl)naphthalene and sulphobenzoic acid anhydride, respectively	PET	[11, 14]
Titration	Quantification of carboxyl groups	PET, PA	[18, 21]
HPLC, LC-MS, other	Quantification of oligomers and monomers released	PET, PA, PAN	Various
Weight loss	Solubilisation of oligomers	PET, PA	Various
Tensile strength	Bulk properties	PET, PA, PAN	Various

oxygen content and to quantify the amount of carboxyl and hydroxyl groups inserted upon enzymatic hydrolysis of PET [2, 13, 16, 17]. Alternatively, carboxyl groups or hydroxyl groups in PET were quantified on the basis of esterification with fluorescent 2-(bromoniethyl)naphthalene (BrNP) or sulphobenzoic acid anhydride, respectively [11, 14]. For the same purpose, other authors have used a titration method quantifying NaOH consumption under constant pH conditions [18]. Differential scanning calorimetry (DSC) and Fourier transform infrared spectroscopy (FTIR) allow determination of crystallinity changes during enzymatic hydrolysis of PET and PTT, which usually occur due to the preference of the enzymes for amorphous regions [15]. FTIR was also used to follow formation of amide and amino groups during hydrolysis of PAN and PA [16, 19]. Quantification of ammonia released from PAN is a simple method and yet reliable for quantifying surface hydrolysis (if proceeding to the corresponding acid) [16]. Likewise, monomeric and oligomeric reaction products released during hydrolysis of PA and PET were determined with simple spectrophotometry (e.g. terephthalic acid) or more sophisticated tools (e.g. LC-MS). The release of molecules from polymers is certainly a proof of enzyme hydrolysis and reveals useful information about specificities of different enzymes. However, a high exo-type hydrolysis activity resulting in the release of water-soluble monomers and short-chain oligomers will lead to polymer degradation rather than functionalisation. Thus, for the study of *endo*-type surface hydrolysis, other techniques such as MALDI-TOF have been used for quantifying the formation of large fragments [2].

2 Polyamidases

Certain proteases, amidases, lipases and cutinases have been shown to hydrolyse polyamide 6 and polyamide 6,6 [20, 22–25]. Consequently, the term polyamidases comprises enzymes from distinct classes but which show the ability to act on water-insoluble polyamides. Initially, the potential of enzymes for PA hydrolysis was assessed both to recover monomers and to degrade oligomers resulting during production. For the first issue, enzymes from *Comamonas acidovorans* were described [26]. On the other hand, enzymes from *Arthrobacter* sp. and *Pseudomonas* sp. have been investigated for the hydrolysis of nylon oligomers [27]. Apart from enzymatic degradation of PA or PA oligomers, targeted (limited) surface hydrolysis has been investigated. The concomitant reduction of PA hydrophobicity has been demonstrated to enhance the wearing comfort of PA fabrics or to prevent fouling of PA-based ultrafiltration membranes [6–9]. Thereby, the enzyme-based process represents an environmentally friendly and highly specific alternative to chemical or plasma treatments [28–30].

Recently, improved hydrophilicity and dyeability with acid and disperse dyes of nylon 6 fibres after treatment with protease (subtilisin) was reported [31]. Similarly, various proteases were used for surface hydrolysis of nylon 6,6 fibres, leading to higher dyebath exhaustion for reactive and acid dyes. PA surface hydrolysis was demonstrated by using FTIR analysis based on changes in the 3300–3500, 1533, 1657 and 1000–1300 cm^{-1} areas [19]. Based on DSC studies, no changes in glass transition and melting points were observed upon limited enzyme hydrolysis. Other authors have quantified the release of monomeric and oligomeric reaction products resulting during hydrolysis of PA with a protease *Bacillus subtilis*. Again, treatment with this enzyme led to increased hydrophilicity and enhanced binding of reactive dyes [22]. Hydrolysis of PA monofilaments with aspartic protease Protease M, cysteine protease Bromelain and metallo-protease Corolase N was confirmed by titration of generated carboxylic end-groups [21].

Apart from proteases, cutinases and lipases were demonstrated to hydrolyse PA. Hydrolysis of PA with a lipase (not further specified) was confirmed by using FTIR analysis and dye-binding assays [32]. Several reports have assessed the potential of cutinases for PA functionalisation, and a fungal cutinase from *F. solani* was successfully genetically engineered towards higher activity on a PA oligomer and PA [23–25].

Production of amidases capable of hydrolysing PA was recently reported both for fungi and bacteria. An amidase from the fungus *Beauveria brongniartii* increased the hydrophilicity of PA6, resulting in a reduction of the drop dissipation time from 60 to 7 s after 60 min incubation, while the surface tension σ increased after 3 min of enzyme treatment from 46.1 to 67.4 mNm [33]. This 55 kDa amidase hydrolysed both aliphatic and aromatic amines but did not show protease activity. Similarly, a bacterial amidase from *Nocardia farcinica* with PA hydrolase activity did not show protease activity [20]. Again, considerable hydrophilicity increases of PA were measured, based on rising height and tensiometry measurements after treatment with this enzyme. The *N. farcinica* polyamidase belongs to the amidase signature family and, consequently, hydrolysed various small amides and esters including *p*-nitroacetanilide and *p*-nitrophenylbutyrate [34].

Unlike serine proteases, lipases and esterases, which are all characterised by the catalytic triad Ser-His-Asp, the catalytic reaction of this polyamidase involves the Ser-Ser-Lys triad [35, 36]. Interestingly, individual representatives of the amidase signature family enzymes show very distinct substrate specificities, which could be due to binding of the substrate by residues outside the signature sequence [35, 36]. Surprisingly, several closely related amidases (based on amino acid sequences) within the amidase signature family did not hydrolyse PA but were reported to degrade cyclic nylon oligomers [37, 38] (Fig. 1). For PA oligomer hydrolysis by *Arthrobacter* sp. KI72 and *Pseudomonas* sp. NK87, a 6-aminohexanoate-cyclic-dimer hydrolase (EI), a 6-aminohexanoate-dimer hydrolase (EII) and an *endo*-type 6-aminohexanoate-oligomer hydrolase (EIII) have been described [27]. EIII hydrolyses the cyclic tetramer and dimer as well as linear oligomers *endo*-wise [37]. Like the *N. farcinica* polyamidase, the *cyclic* dimer hydrolases belonged to the amidase signature family whereas the *linear* dimer hydrolase (EII) activity evolved in an esterase with β-lactamase folds. In contrast, the *endo*-acting EIII showed the lowest homology to the *N. farcinica* polyamidase [20, 27].

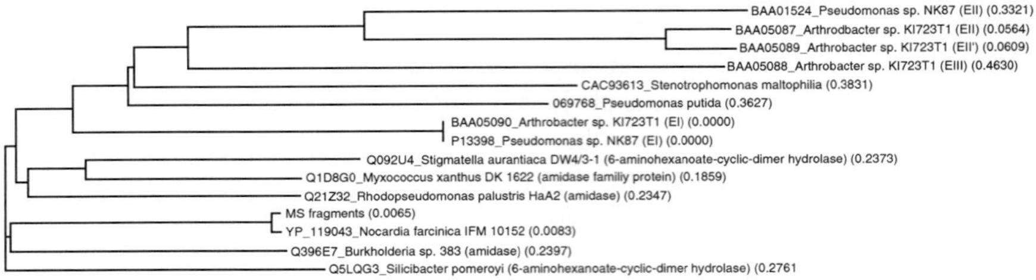

Fig. 1 Phylogenetic tree based on amino acid sequences of the polyamidase from *Nocardia farcinica* and other highly homologous bacterial amidases as well as amidases with substrate specificity for 6-aminohexanoate oligomers [20]

3 PET Hydrolases

As for PA-hydrolysing enzymes, representatives from different enzyme classes including lipases, esterases, cutinases and proteases have been demonstrated to hydrolyse PET and are termed PET hydrolases here. Searching for PET hydrolases in nature, the first choice was the investigation of enzymes that hydrolyse the hydrophobic plant polyester cutin. Cutin from the plant cuticle consists of oxygenated C16 and C18 fatty acids crosslinked by ester bonds [39] and is essential for plant protection. Cutin degradation by cutinases is one of the first steps in the infection of plants [40]. Cutinases show both *exo-* and *endo-*esterase activity [41] and have first been investigated from *F. solani* pisi growing on cutin as a carbon source [42]. Cutin oligomers have been suggested to induce production of these enzymes [43].

Interestingly, some PET-hydrolyzing enzyme activities were likewise inducible by addition of cutin [44, 45]. A number of fungal cutinases, such as from *F. solani* and *Fusarium oxysporum* [13, 46, 47]; *Aspergillus oryzae*, *Aspergillus niger* and *Aspergillus nomius* [48–50]; *Humicola* sp. and *Humicola insolens* [47, 51, 52]; and from *Penicillium citrinum* [45], have shown PET hydrolase activity. Cutinases carry their active site at the surface of the protein, which is essential for *endo-*wise hydrolysis of polymers. Recently, a comparison of the substrate specificities and structures of the *F. solani* and *A. oryzae* cutinases revealed a preference of the latter enzyme to hydrolyse longer chain substrates. This is probably due to a deep continuous groove extending across the active site, in contrast to that of *F. solani* with a shallow and interrupted groove (Fig. 2). Consequently, the *A. oryzae* cutinase showed higher activity on the polyester poly(ε-caprolactone) [53]. Cutinases are also widespread in bacteria, and representatives from *Thermobifida fusca* [2, 15, 44, 54], *Thermobifida alba* [52] and *Pseudomonas mendocina* [18] have shown PET hydrolase activity.

In addition to cutinases, various lipases, such as from *C. antarctica*, *Candida* sp. [13, 47], *Thermomyces lanuginosus* [2, 14, 15, 55, 56], *Burkholderia* (formerly *Pseudomonas*) *cepacia* [57] and esterases from *Pseudomonas* sp. (serine esterase) [58] and *Bacillus* sp. (nitrobenzyl esterases) [59, 60], have shown PET hydrolase

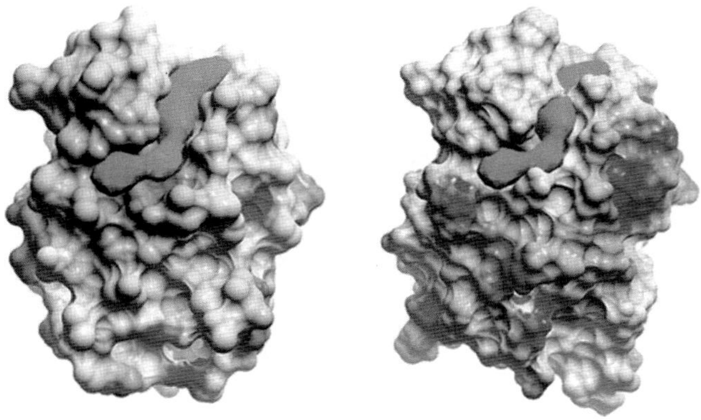

Fig. 2 Electrostatic surface rendering of *Aspergillus oryzae* (*left*) and *Fusarium solani* (*right*) cutinases. The *solid density* illustrates the groove on the surface proximal to the active site [53]

activity. Unlike cutinases, lipases are "interfacially activated" in the presence of a water–lipid interface [61–63]. Consequently PET hydrolase activity was enhanced in the presence of detergents [2, 64].

Several studies have reported on hydrophilicity increases of PET after limited enzyme hydrolysis [14, 46, 47, 54, 65]. For instance, hydrolysis of PET fabrics with cutinases or lipases resulted in a wetting time of around 100 s compared to 45 min for the untreated material [1] and in considerable decreases of the WCA [14]. Derivatisation with BrNP [14] or sulphobenzoic acid anhydride [11] indicated a concomitant increase in the number of carboxyl and hydroxyl groups, respectively. XPS analysis yielded the same message, based on broader carbon peaks after enzyme treatment of PET [2, 13]. Likewise, enzyme hydrolysis of PET was followed by using dye-binding assays, with basic dyes resulting in colour shade increases of up to 200% (according to the Kubelka Monk theory) [1, 54, 60]. Limited surface hydrolysis did not reveal morphological changes according to SEM inspections, except for prolonged incubation of low crystallinity PET (Fig. 3). In contrast to the considerable weight losses (>6% for 1 M NaOH) and crater-like structures (Fig. 3) reported after alkaline hydrolysis, most authors reported only insignificant weight losses ($\ll$1%) for enzyme hydrolysis yet obtained similar hydrophilicity increases [1, 14]. Obviously, this is due to the fact that (successful) enzymes hydrolyse PET *endo*-wise. Only recently was this demonstrated for PET ($M_W = 3500$) by using MALDI-TOF MS, whereas a different behaviour for cutinases and lipases was seen [2]. Nevertheless, in several studies the release of mono- and oligomeric reaction products from PET hydrolysis was demonstrated [13, 45–47, 49, 54, 55, 58]. For example, a lipase from *T. lanuginosus* released higher amounts of mono(2-hydroxyethyl) terephthalate (MHET) than of terephthalic acid (TA), whereas the amounts of TA and MHET were similar in the case of a cutinase from *T. fusca*. Small amounts of bis(2-hydroxyethyl) terephthalate (BHET) were detected for both enzymes [1].

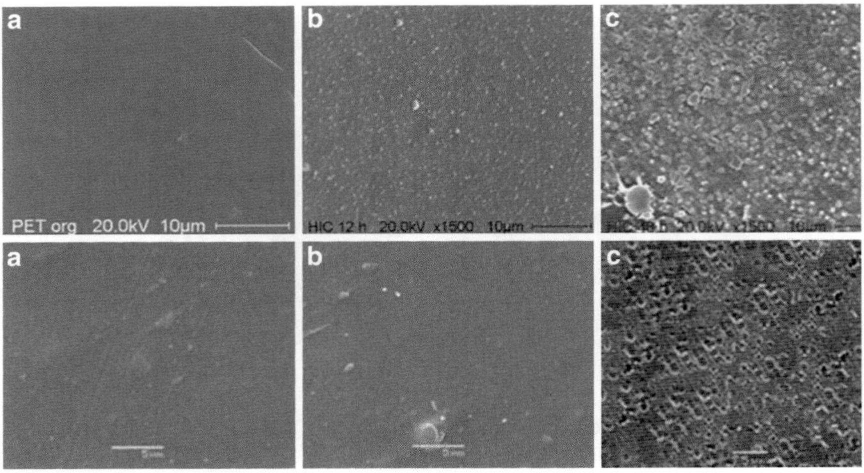

Fig. 3 *Top row*: Hydrolysis of PET with 7% crystallinity with a *Humicola insolens* cutinase for (**a**) 0, (**b**) 12 and (**c**) 48 h resulting in weight losses of 0, 18% and 54%, respectively. *Bottom row*: Hydrolysis of PET with (**a**) 34.8% crystallinity (blank), (**b**) a lipolytic enzyme (no weight loss) and (**c**) 1 M NaOH (modified from [14, 18])

Several methods, including DSC and FTIR, have been used to demonstrate increases of crystallinity during hydrolysis of PET [14, 18]. This indicates that PET hydrolases preferably attack the amorphous regions of PET [13–15, 44, 66, 67]. Indeed, a cutinase from *T. fusca* released up to 50-fold higher amounts of oligomers and terephthalic acid from amorphous fibres than from semi-crystalline fibres [1]. Likewise, a lipase displayed higher hydrolytic activity towards amorphous PET, as shown by the decrease in the WCA values [14]. In the same way, NaOH titration at constant pH indicated a tenfold increase in activity of cutinases on PET with 7% crystallinity compared to PET with 35% crystallinity [18]. Similar results were obtained with PTT (polytrimethyleneterephthalate), which is gaining increasing importance because, apart from attractive properties, one of the building blocks (namely 1,3-propanediol) can be produced by microbial fermentation from renewable sources [15].

4 Polyacrylonitrilases

PAN was for a long time thought to be resistant to microbial attack. However, various bacteria that produced nitrile-converting enzymes were isolated from waste-waters of factories producing PAN fibre. For example, a nitrile hydratase/amidase enzyme system was studied from *Mesorhizobium* sp. F28 [68]. Also, bacteria (namely *Ralstonia solanacearum* and *Acidovorax avenae*) were used for the removal of acrylic acid from such waste-waters [69]. Later, on the basis of NMR analysis of the release of polyacrylic acid and on microscopic inspection, bacteria such as *Micrococcus luteus* were demonstrated to degrade PAN [5].

Generally, enzymatic hydrolysis of nitriles to the corresponding acids can either proceed stepwise, which is the case for catalysis by the nitrile hydratase/amidase enzyme system, or in one step in the case of nitrilases. Both systems have been investigated for surface hydrolysis of PAN [10]. Complete hydrolysis with either system was monitored by quantification of ammonia and/or polyacrylic acid formed as a consequence of hydrolysis of nitrile groups [70–72]. As a result, considerable increases in colour levels (e.g. 156% for commercial nitrilase) were found upon dyeing [72].

XPS analysis and FTIR were used to quantify chemical changes on the surface of PAN materials [5, 16]. For PAN treated with nitrilases from *Arthrobacter* sp. and *Agrobacterium tumefaciens*, increases of the O/C ratio of 60–80%, respectively, were measured with XPS [17]. The conversion of nitrile groups into amide groups was demonstrated on the basis of formation of new bands at 1649 and 1529 cm^{-1} in FTIR analysis [16]. Generally, as with PET and PA, lower hydrolysis rates were measured for crystalline PAN [5, 16]. Interestingly, depending on the enzyme system used, either hydrolysis to the corresponding amides or further hydrolysis to the acids can be achieved. Nitrile hydratases from *Rhodococcus rhodochrous*, *Brevibacterium imperiale* and *Corynebacterium nitrilophilus* lead to the corresponding amides, whereas further hydrolysis by the amidase seems to be slower [16, 71, 73].

Apart from nitrile-hydrolyzing enzymes, some esterases and cutinases have been used for surface hydrolysis of PAN [74]. These enzyme were shown to specifically hydrolyse vinyl acetate moieties present as co-monomer in many commercial PAN materials, with no changes in crystallinity as determined by X-ray diffraction [74].

Summarising the recent progress in enzymatic surface hydrolysis of PA, PET and PAN, recent detailed mechanistic studies using distinct analytical tools have contributed considerably to our mechanistic understanding of the enzyme action of these polymers. Investigations of structure–function relationships and enzyme homologies yield important information as a basis for enzyme engineering and screening for more efficient enzymes.

Acknowledgement The work was financed by the SFG, the FFG, the city of Graz and the province of Styria within the MacroFun project and supported by the European COST868 program.

References

1. Brueckner T, Eberl A, Heumann S et al (2008) Enzymatic and chemical hydrolysis of poly(ethylene terephthalate) fabrics. J Polym Sci 46:6435–6443
2. Eberl A, Heumann S, Brueckner T et al (2009) Enzymatic surface hydrolysis of poly(ethylene terephthalate) and bis(benzoyloxyethyl) terephthalate by lipase and cutinase in the presence of surface active molecules. J Biotechnol 143:207–212
3. Zeronian SH, Collins MJ (1989) Surface modification of polyester by alkaline treatments. Textil Progr 20:1–34

4. Hsieh YL, Miller A, Thompson J (1996) Wetting, pore structure and liquid retention of hydrolyzed polyester fabrics. Textil Res J 66:1–10
5. Fischer-Colbrie G, Matama T, Heumann S et al (2007) Surface hydrolysis of polyacrylonitrile with a nitrilase of a new strain of *Micrococcus luteus*. J Biotechnol 128:849–857
6. Asatekin A, Kang S, Elimelech M et al (2007) Anti-fouling ultrafiltration membranes containing polyacrylonitrile-graft-poly(ethylene oxide) comb copolymer additives. J Membr Sci 298:136–146
7. Kim HA, Choi JH, Takizawa S (2007) Comparison of initial filtration resistance by pretreatment processes in the nanofiltration for drinking water treatment. Separ Purif Technol 56:354–362
8. Qiao X, Zhang Z, Ping Z (2007) Hydrophilic modification of ultrafiltration membranes and their application in *Salvia miltiorrhiza* decoction. Separ Purif Technol 56:265–269
9. Li JX, Wang J, Shen LR et al (2007) The influence of polyethylene terephthalate surfaces modified by silver ion implantation on bacterial adhesion behavior. Surf Coat Technol 201:8155–8159
10. Guebitz GM, Cavaco-Paulo A (2008) Enzymes go big: surface hydrolysis and functionalisation of synthetic polymers. Trends Biotechnol 26:32–38
11. Almansa E, Heumann S, Eberl A et al (2008) Enzymatic surface hydrolysis of PET enhances bonding in PVC coating. Biocatal Biotrans 26:365–370
12. Laskarakis A, Logothetidis S, Kassavetis S et al (2007) Surface modification of poly(ethylene terephthalate) polymeric films for flexible electronics applications. Thin solid films 516: 1443–1448
13. Vertommen MAME, Nierstrasz VA, Veer Mvd et al (2005) Enzymatic surface modification of poly(ethylene terephthalate). J Biotechnol 120:376–386
14. Donelli M, Taddei P, Smet P F et al (2009) Enzymatic surface modification and functionalization of PET. A water contact angle, FTIR, and fluorescence spectroscopy study. Biotechnol Bioeng 103:845–856
15. Eberl A, Heumann S, Kotek R. et al (2008) Enzymatic hydrolysis of PTT polymers and oligomers. J Biotechnol 135:45–51
16. Fischer-Colbrie G, Herrmann M, Heumann S et al (2006) Surface modification of polyacrylonitrile with nitrile hydratase and amidase from *Agrobacterium tumefaciens*. Biocatal Biotrans 24:419–425
17. Wang N, Xu Y, Lu D (2004) Enzymatic surface modification of acrylic fiber. AATCC Rev 4:28–30
18. Ronkvist AM, Xie WC, Lu WH et al (2009) Cutinase-catalyzed hydrolysis of poly(ethylene terephthalate). Macromolecules 42:5128–5138
19. Parvinzadeh M, Assefipour R, Kiumarsi A (2009) Biohydrolysis of nylon 6,6 fiber with different proteolytic enzymes. Polym Degrad Stab 94:1197–1205
20. Heumann S, Eberl A, Fischer-Colbrie G et al (2009) A novel aryl acylamidase from *Nocardia farcinica* hydrolyses polyamide. Biotechnol Bioeng 102:1003–1011
21. Miettinen-Oinonen A, Puolakka A, Buchert J (2007) Method for modifying polyamide. Patent EP1761670 Finland
22. Silva C, Araujo R, Casal M et al (2007) Influence of mechanical agitation on cutinases and protease activity towards polyamide substrates. Enzyme Microb Technol 40:1678–1685
23. Silva C, Carneiro F, O'Neill A et al (2005) Cutinase – a new tool for biomodification of synthetic fibers. J Polym Sci 43:2448–2450
24. Araujo R, Silva C, O'Neill A et al (2007) Tailoring cutinase activity towards polyethylene terephthalate and polyamide 6,6 fibers. J Biotechnol 128:849–857
25. Silva C, Matama T, Guebitz G M et al (2005) Influence of organic solvents on cutinase stability and accessibility to polyamide fibers. J Polym Sci A Polym Chem 43:2749–2753
26. Crouzet J, Faucher D, Favre-Bovine G, Jourdat C, Petre D, Pierrard J, Thibault J, Guitton C Enzymes and micro organisms with amidase activity which hydrolyze polyamides. US Patent 6180388
27. Negoro S, Ohki T, Shibata N et al (2007) Nylon-oligomer degrading enzyme/substrate complex: catalytic mechanism of 6-aminohexanoate-dimer hydrolase. J Mol Biol 370:142–156

28. De Geyter N, Morent R, Leys C et al (2007) Treatment of polymer films with a dielectric barrier discharge in air, helium and argon at medium pressure. Surf Coat Technol 201:7066–7075
29. McCord MG, Hwang YJ, Hauser PJ et al (2002) Modifying nylon and polypropylene fabrics with atmospheric pressure plasmas. Textil Res J 72:491–498
30. Tusek L, Nitschke M, Werber C et al (2001) Surface characterization of NH_3 plasma treated polyamide foils. Colloid Surf. A: Physicochem. Eng. Aspect 195:81–95
31. Parvinzadeh M (2009) A new approach to improve dyeability of nylon 6 fibre using a subtilisin enzyme. Coloration Technol 125:228–233
32. Parvinzadeh M, Kiumarsi A (2010) Lipase enzyme to improve dyeability of polyamide substrate. J Biotechnol 136:299
33. Almansa E, Heumann S, Eberl A et al (2008) Surface hydrolysis of polyamide with a new polyamidase from *Beauveria brongniartii*. Biocatal Biotrans 26:371–377
34. Yoshioka H, Nagasawa T, Yamada H (1991) Purification and characterization of aryl acylamidase from *Nocardia globerula*. Eur J Biochem 199:17–24
35. Labahn J, Neumann S, Buldt G et al (2002) An alternative mechanism for amidase signature enzymes. J Mol Biol 322:1053–1064
36. Valina ALB, Mazumder-Shivakumar D, Bruice T C (2004) Probing the Ser-Ser-Lys catalytic triad mechanism of peptide amidase: computational studies of the ground state, transition state, and intermediate. Biochemistry 43:15657–15672
37. Kakudo S, Negoro S, Urabe I et al (2000) Nylon oligomer degradation gene, nylC, on plasmid pOAD2 from a *Flavobacterium* strain encodes endo-type 6-aminohexanoate oligomer. Appl Environ Microbiol 59:3978–3980
38. Negoro S (2000) Biodegradation of nylon oligomers. Appl Microbiol Biotechnol 54:461–466
39. Fett WF, Wijey C, Moreau RA et al (1998) Production of cutinase by *Thermomonospora fusca* ATCC 27730. J Appl Microbiol 86:561–568
40. Carvalho CML, Aires-Barros MR, Cabral JMS (1998) Cutinase structure, function and biocatalytic applications. Electron J Biotechnol 1:160–173
41. Kolattukudy PE (1981) Structure, biosynthesis, and biodegradation of cutin and suberin. Annu Rev Plant Physiol Plant Mol Biol 32:539–567
42. Kolattukudy PE, Rogers LM, Li DX et al (1995) Surface signaling in pathogenesis. Proc Natl Acad Sci USA 92:4080–4087
43. Lin TS, Kolattukudy PE (1978) Induction of a biopolyester hydrolase (cutinase) by low levels of cutin monomers in *Fusarium solani* f. sp. pisi. J Bacteriol 133:942–951
44. Mueller RJ (2006) Biological degradation of synthetic polyesters – enzymes as potential catalysts for polyester recycling. Process Biochem 41:2124–2128
45. Liebminger S, Eberl A, Sousa F et al (2007) Hydrolysis of PET and bis-(benzoyloxyethyl) terephthalate with a new polyesterase from *Penicillium citrinum*. Biocatal Biotrans 25: 171–177
46. Alisch-Mark M, Herrmann A, Zimmermann W (2006) Increase of the hydrophilicity of polyethylene terephthalate fibres by hydrolases from *Thermomonospora fusca* and *Fusarium solani* f. sp. pisi. Biotechnol Lett 28:681–685
47. Nimchua T, Punnapayak H, Zimmermann W (2007) Comparison of the hydrolysis of polyethylene terephthalate fibers by a hydrolase from *Fusarium oxysporum* LCH I and *Fusarium solani f. sp. pisi.* Biotechnol J 2:361–364
48. Gouveia I, Queiroz J, Antunes L (2009) Improving surface energy and hydrophilization of poly(ethylene terephthalate) by enzymatic treatments. In: Freire Bastos T, Gamboa H (eds) Biodevices 2009. INSTICC Press, Setúbal
49. Wang X, Lu D, Jonsson LJ et al (2008) Preparation of a PET-hydrolyzing lipase from *Aspergillus oryzae* by the addition of bis(2-hydroxyethyl) terephthalate to the culture medium and enzymatic modification of PET fabrics. Eng Life Sci 8:268–276
50. Uchida H, Kurakata Y, Sawamura H et al (2003) Purification and properties of an esterase from *Aspergillus nomius* HS-1 degrading ethylene glycol dibenzoate. FEMS Microbiol Lett 223:123–127
51. Liu YB, Wu GF, Gu LH (2008) Enzymatic treatment of PET fabrics for improved hydrophilicity. AATCC Rev 8:44–48

52. Korpecka J (2009) Cutinase activity of PET-hydrolases. In: Proceedings of INTB 2009, Ghent, Belgium, September 2009
53. Liu ZQ, Gosser Y, Baker PJ et al (2009) Structural and functional studies of *Aspergillus oryzae* cutinase: enhanced thermostability and hydrolytic activity of synthetic ester and polyester degradation. J Am Chem Soc 131:15711–15716
54. Alisch M, Feuerhack A, Mueller H et al (2004) Biocatalytic modification of polyethylene terephthalate fibres by esterases from actinomycete isolates. Biocatal Biotrans 22:347–351
55. Andersen BK, Borch K, Abo M et al (1999) Method of treating polyester fabrics. US Patent 5,997,584, pp 1–20
56. Heumann S, Eberl A, Pobeheim H et al (2006) New model substrates for enzymes hydrolysing polyethyleneterephthalate and polyamide fibres. J Biochem Biophys Methods 69:89–99
57. Lee CW, Do Chung J (2009) Synthesis and biodegradation behavior of poly(ethylene terephthalate) oligomers. Polymer (Korea) 33:198–202
58. Yoon MY, Kellis J, Poulouse AJ (2002) Enzymatic modification of polyester. AATCC Rev 2:33–36
59. Michels A, Pütz A, Maurer KH, Eggert T, Jäger K-E Use of esterases for separating plastics. WO/2007/017181 Germany
60. Kellis J, Poulose AJ, Yoon MY Enzymatic modification of the surface of a polyester fiber or article. US Patent 6,254,645 B1, US 6,254,645 B1
61. Grochulski P, Li Y, Schrag JD et al (1993) Insights into interacial activation from an open structure of *Candida rugosa* Lipase. J Biol Chem 286:12843–12847
62. Fojan P, Jonson PH, Petersen MTN et al (2000) What distinguishes an esterase from a lipase: a novel structural approach. Biochimie 82:1033–1041
63. Pleiss J, Fischer M, Schmid RD (1998) Anatomy of lipase binding sites: the scissile fatty acid binding site. Chem Phys Lipids 93:67–80
64. Kim HR, Song WS (2006) Lipase treatment of polyester fabrics. Fibers Polym 7:339–343
65. Fischer-Colbrie G, Heumann S, Liebminger S et al (2004) New enzymes with potential for PET surface modification. Biocatal Biotrans 22:341–346
66. Herzog K, Müller RJ, Deckwer WD (2006) Mechanism and kinetics of the enzymatic hydrolysis of polyester nanoparticles by lipases. Polym Degrad Stab 91:2486–2498
67. Müller RJ, Schrader H, Profe J et al (2005) Enzymatic degradation of poly(ethylene terephthalate): Rapid hydrolyse using a hydrolase from *T. fusca.* Macromol Rapid Commun 26:1400–1405
68. Feng YS, Chen PC, Wen FS et al (2008) Nitrile hydratase from *Mesorhizobium* sp F28 and its potential for nitrile biotransformation. Process Biochem 43:1391–1397
69. Wang CC, Lee CM, Wu AS (2009) Acrylic acid removal from synthetic wastewater and industrial wastewater using *Ralstonia solanacearum* and *Acidovorax avenae* isolated from a wastewater treatment system manufactured with polyacrylonitrile fiber. Water Sci Technol 60:3011–3016
70. Tauber MM, Cavaco-Paulo A, Gübitz GM (2001) Enzymatic treatment of acrylic fibers and granulates. AATCC Rev 1:17–19
71. Tauber MM, Cavaco-Paulo A, Robra K-H et al (2000) Nitrile hydratase and amidase from *Rhodococcus rhodochrous* hydrolyse acrylic fibers and granulates. Appl Environ Microbiol 66:1634–1638
72. Matama T, Carneiro F, Caparrós C et al (2007) Using a nitrilase for the surface modification of acrylic fibres. Biotechnol J 2:353–360
73. Battistel E, Morra M, Marinetti M (2001) Enzymatic surface modification of acrylonitrile fibers. Appl Surface Sci 177:32–41
74. Matama T, Vaz F, Gubitz GM et al (2006) The effect of additives and mechanical agitation in surface modification of acrylic fibres by cutinase and esterase. Biotechnol J 1:842–849

Adv Polym Sci (2010) 237: 127–143
DOI:10.1007/12_2010_75

Published online: 7 September 2010

Exploiting Biocatalysis in the Synthesis of Supramolecular Polymers

Sangita Roy and Rein V. Ulijn

Abstract This chapter details the exploitation of biocatalysis in generating supramolecular polymers. This approach provides highly dynamic supramolecular structures, inspired by biological polymeric systems found in the intra- and extracellular space. The molecular design of the self-assembling precursors is discussed in terms of enzyme recognition, molecular switching mechanisms and non-covalent interactions that drive the supramolecular polymerisation process, with an emphasis on aromatic peptide amphiphiles. We discuss a number of unique features of these systems, including spatiotemporal control of nucleation and growth of supramolecular polymers and the possibility of kinetically controlling mechanical properties. Fully reversible systems that operate under thermodynamic control allow for defect correction and selection of the most stable structures from mixtures of monomers. Finally, a number of potential applications of enzymatic supramolecular polymerisations are discussed in the context of biomedicine and nanotechnology.

Keywords Biocatalysis · Enzymes · Peptides · Molecular hydrogel · Self-assembly · Supramolecular polymers

Contents

S. Roy and R.V. Ulijn (✉)
WestCHEM, Department of Pure and Applied Chemistry, University of Strathclyde, 295 Cathedral Street, Glasgow, G1 1XL UK
e-mail: rein.ulijn@strath.ac.uk

1 Introduction

During the last two decades, chemists have become increasingly focused on how molecules interact, i.e. on supramolecular chemistry. Dynamic intermolecular processes provide opportunities for incorporation of control, adaptation and function in man-made materials, as observed in living systems. In biology, these processes are tightly controlled by the catalytic action of enzymes. In this chapter, we focus on enzymatically controlled supramolecular polymerisation, whereby self-recognising molecular building blocks assemble to form extended one-dimensional (1D) structures, or supramolecular polymers, with unique adaptive features.

While traditional chemistry focuses on covalent bonds, supramolecular chemistry examines reversible, non-covalent interactions including hydrogen bonding, metal coordination, hydrophobic interactions, van der Waals forces, π-stacking and electrostatic interactions [1, 2]. Supramolecular polymerisation capitalises on these interactions and involves the controlled association of amphiphilic molecular building blocks into organised structures such as fibres. At high densities, these fibres may become entangled, forming gel-phase structures, and building blocks are therefore often referred to as gelators. Supramolecular polymerisation relies on a precise balance between hydrophobic interactions and interactions of building blocks with solvent, and can be induced by chemical and physical triggers that cause subtle changes in their amphiphilic balance [3]. These triggers include changes in pH [4], ionic strength [5, 6], solvent polarity or temperature, or locally induced chemical transformations catalysed by light [7–9] or (bio)catalysts [10]. Biocatalytically controlled supramolecular polymerisation has a number of attractive features that will be explored in this chapter [11].

We will first discuss some examples of biocatalytic polymerisation in biological systems, followed by a review of recent man-made systems and the design rules that are emerging. Unique features related to control of polymerisation (both in terms of kinetics and thermodynamics) will be discussed, followed by a review of (potential) applications in biomedicine and nanotechnology.

2 Examples from the Biological World

Inspiration for development of dynamic supramolecular polymers comes from living systems, where enzyme-controlled formation and degradation of collagen fibrils, actin filaments and microtubules underlie vital cellular functions such as motility, differentiation, division, etc. (Fig. 1).

Collagen fibres (Fig. 1a) give rise to stiff gel-like materials that surround cells to provide a scaffold to support tissue formation. These gels are formed by entanglement of networks of collagen fibres. Non-assembling precursors known as procollagens are firstly produced within cells. These may form triple helical structures. Further assembly is prevented within cells due to peptide sequences at both ends of the collagen molecules. After secretion into the extracellular space, procollagens are converted to self-assembling collagens by specific enzymatic cleavage of both the N- and C-terminal polypeptides in a sequential fashion to generate collagen fibrils. This mechanism provides exquisite spatial and temporal control of collagen fibril production (Fig. 1a) [12–16]. Certain proteolytic enzymes such as matrix metalloproteases (including collagenase) are capable of selectively degrading collagen fibres, giving rise to a highly dynamic and adaptive system.

Biocatalytic control of supramolecular polymerisation also underlies the dynamics of actin filaments (F-actin) that drive the locomotion of single-cell organisms

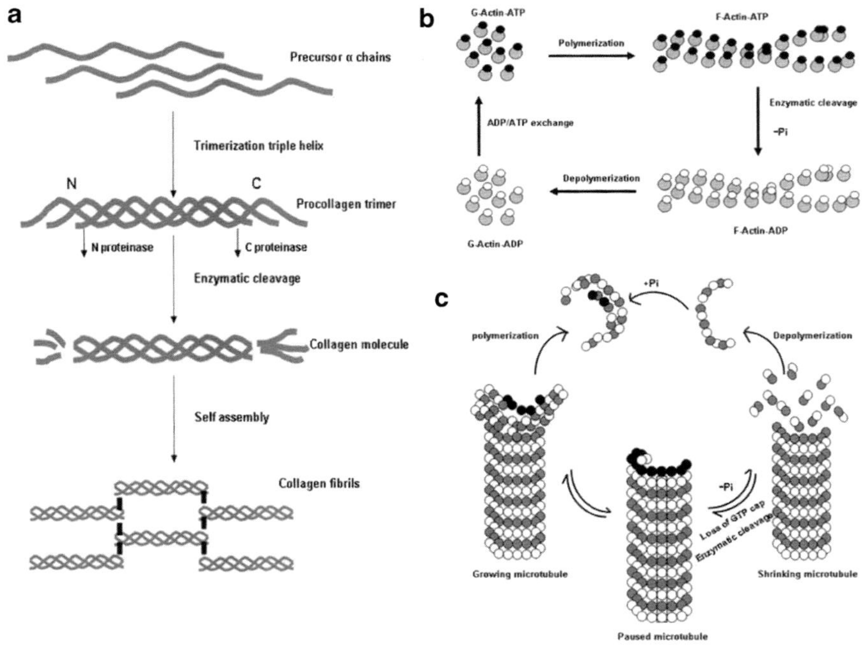

Fig. 1 Examples of enzyme-controlled supramolecular polymerisation from the biological world: (**a**) formation of collagen fibrils, (**b**) dynamic self-assembly of actin filaments, and (**c**) formation of microtubules

[17] as well as muscle contraction in higher organisms. Actin filaments are linear polymers of the monomeric subunit globular actin (G-actin), which undergo continuous polymerisation and depolymerisation. This process is initiated and controlled by the enzymatic hydrolysis of the biomolecular fuel molecule adenosine triphosphate (ATP) to the diphosphate analogue (ADP) by ATPase (Fig. 1b). Likewise, the on-demand formation of microtubules, which are responsible for intracellular transport and cell division during mitosis, is regulated by enzymatic reactions. Here, the enzymatic hydrolysis of guanosine triphosphate (GTP) to GDP by GTPase controls the corresponding assembly and dis-assembly of the monomeric proteins in microtubules (Fig. 1c) [17]. In summary, Nature has evolved a number of dynamic polymeric structures that can form and dissociate under enzymatic control. These play key roles as dynamic intra- and extracellular structures that underlie dynamic cell function. Scientists are interested in copying these highly dynamic processes in man-made systems, as detailed in the following sections.

3 Enzymatic Supramolecular Polymerisation

Building blocks (monomers) for enzyme-controlled supramolecular polymerisations are comprised of three components: (1) an enzyme-specific target (biomolecule based on the enzyme's substrate specificity), (2) a self-assembly component that directs the non-covalent interaction responsible for supramolecular polymerisation, and (3) a molecular switch component that prevents precursor self-assembly and activates self-assembly upon enzyme action.

3.1 Enzyme-Specific Target

A range of enzyme-catalysed reactions have been exploited to control formation of supramolecular polymeric structures [11, 18]. These include enzymatic crosslinking strategies using protein crosslinking enzymes such as transglutaminase [19, 20], enzymatic production of gelators from non-gelling precursors using protease [21–24], esterase [22], penicillin G amidase [25], trans-acylase [26] and phosphatase [27–30] activities. The opposite process, i.e. enzymatic degradation or disassembly of supramolecular structures, has been described in the context of degradable peptide nanotubes [31–33], combined with enzymatically or thermally controlled release [34], and cell culture matrices [35, 36].

Dynamic systems that exploit both assembly and dis-assembly have been explored using reversible (de)phosphorylation in response to kinases (phosphorylation enzymes) and phosphatases [28, 37]. In these systems, the phosphorylation reaction is facilitated by simultaneous hydrolysis of ATP, akin to biological systems discussed in the previous section. The concentration of ATP can dictate the preferred direction of these reactions, with phosphorylation levels (and therefore supramolecular polymerisation) dictated by ATP levels when both kinase and phosphatase are present. Dephosphorylation has also been exploited to control supramolecular association of peptide–polymer conjugates [29, 30]. Although most examples that have been reported to date are based on peptides, enzyme-responsive DNA-based hydrogels have also been described [38] and Nature's designs will continue to inspire supramolecular polymerisation [39–41].

3.2 Self-Assembly Component

Peptides and their derivatives are particularly attractive building blocks for supramolecular polymerisations because of their rich opportunities for non-covalent interactions (hydrogen bonding, electrostatic, π-stacking, hydrophobic), ease of synthesis and compatibility with biological systems. There are 20 gene-coding amino acids commonly found in nature (identified in this text by their single-letter codes, Fig. 2a) plus an infinite number of modified and non-natural monomers. The properties and self-assembly tendency of peptides generated from the combination of these amino acids depends on the nature of their side-chain substitution. Although an enormous number of possible sequences exist (20^5; 3.2 million sequences for

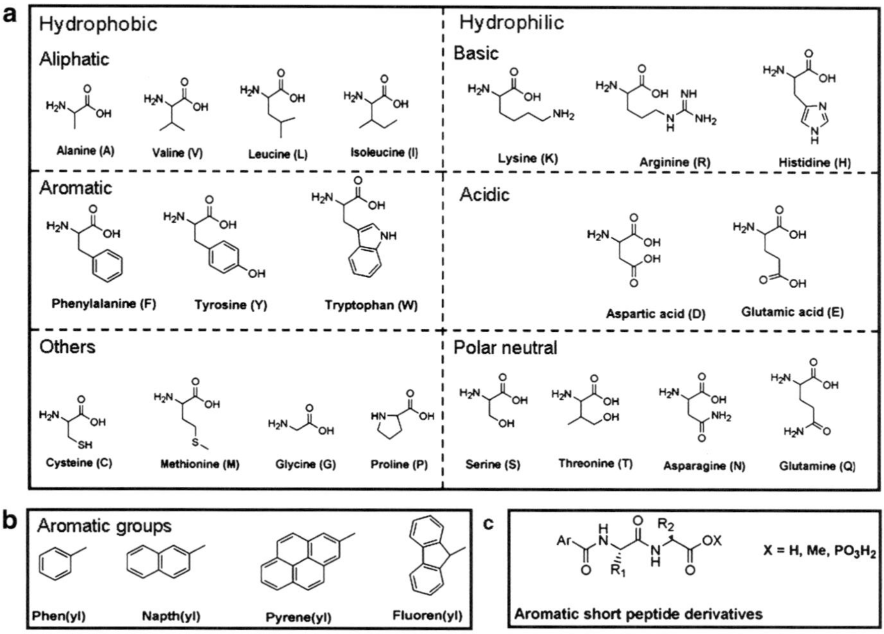

Fig. 2 The structure of (**a**) 20 natural L amino acids and (**b**) aromatic residues that gives rise to (**c**) aromatic peptide amphiphiles, which form supramolecular polymers through hydrogen bonding and π-stacking interactions

a pentapeptide), only a few of them are found in biological systems. Considerable progress has been made towards elucidating design rules for peptide-based supramolecular materials [42–45]. The design rules are either derived by copying nature (α-helix, β-sheet) [46, 47] or are entirely new designs that exploit peptide derivatives such as aliphatic [11, 48] or aromatic peptide amphiphiles [10, 49–52]. The latter systems allow for the use of much simpler, shorter peptides (as small as dimers) compared to other approaches that usually require at least ten amino acids (and often many more) in each peptide chain. This approach facilitates rational design and lowers costs for eventual application.

We focus on enzymatically controlled supramolecular polymerisations based on aromatic peptide amphiphiles as building blocks because these are by far the most widely studied systems. These are short peptides (generally one to five amino acids) that are modified (usually at the N-terminus) with aromatic groups such as phenyl, napthyl, pyrene, 9-fluorenyl methoxycarbonyl (Fmoc) etc. (Fig. 2b, c). It was first highlighted in the mid-1990s that certain Fmoc-dipeptides can self-assemble to form gel-phase materials [53]. For a number of aromatic peptide amphiphiles, it has been shown that self-assembly is governed by formation of in-register antiparallel β-sheet structures with aromatic groups extended at both termini of each sheet. Multiple sheets were found to lock together via (antiparallel) π-stacking interactions to give rise to π-interlocked β-sheets or π–β structures (Fig. 3) [51, 54]. The nature of the amino acid residues in these systems dictates the curvature of the β-sheets, and in cases where curvature allows both edges of an array of β-sheets to lock together fibres or hollow tubes may form [21, 22]. It is clear that the morphologies and the dimensions of these nanostructures are strongly dependent on the route of self-assembly (see Sect. 3.3) as well as on the amino acid sequence and the chemical nature of the aromatic residues in the building blocks.

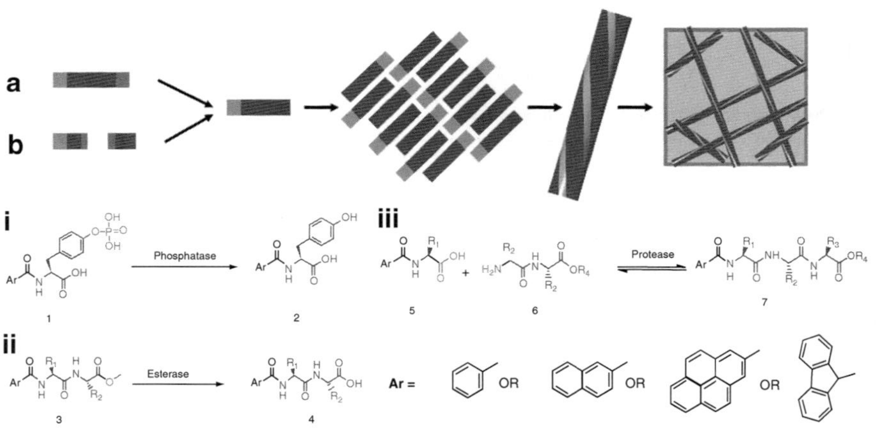

Fig. 3 Mechanisms for enzymatic supramolecular polymerisation: (**a**) Formation of supramolecular assembly via bond cleavage. (**b**) Formation of supramolecular assemblies via bond formation. Examples are shown of biocatalytic supramolecular polymerisation of aromatic peptide amphiphiles via (i) phosphate ester hydrolysis, (ii) alkyl ester hydrolysis, and (iii) amide condensation or reversed hydrolysis using protease

3.3 Molecular Switch Component

A number of routes have been explored in enzymatic activation of precursors to self-assembly building blocks, either exploring bond cleavage (hydrolysis) or bond formation (Fig. 3).

Building blocks are amphiphiles, which have a delicate balance between the hydrophilic and hydrophobic group crucial to facilitate self-assembly. The peptide component serves to precisely control this balance, and the enzymatic reaction serves to alter it in favour of self-assembly. As illustrated in Fig. 3, the molecular switch may involve: (1) phosphatase-catalysed removal of a (phosphate) group from the precursor to control the electrostatic balance (reaction (i) in Fig. 3); (2) hydrolysis of alkyl esters by hydrolases to change the amphiphilic balance (reaction (ii) in Fig. 3); or (3) condensation between two non-self-assembling precursors via a condensation reaction, e.g. involving protease-catalysed amide synthesis to alter the hydrophilic/hydrophobic balance (reaction (iii) in Fig. 3). A number of examples of each type are summarised in Table 1.

In addition to chemical composition, as discussed in Sect. 3, the route of self-assembly also significantly affects the resulting structure because different kinetically folded structures may be formed. For example, subtilisin-triggered formation of Fmoc-L_3 via ester hydrolysis gives rise to hollow nanotubular structures [22], whereas Fmoc-L_3 gel formed by the thermolysin-catalysed reversed hydrolysis of the Fmoc-L/L_2 system gives rise to nanofibrillar morphology [21].

Table 1 Examples of enzymatic formation of supramolecular assemblies

Enzymes	Enzyme-sensitive component	Component forming molecular self-assembly	Mode of formation	References
Alkaline phosphatase	(a) Fmoc-*p*Y (phosphate ester)	(a) Fmoc-Y	Via bond cleavage	[10, 27, 37, 55]
	(b) Nap-FFGE*p*Y	(b) Nap-FFGEY		
Subtilisin	Fmoc-$(L)_2$OMe, Fmoc-$(L)_3$OMe, Fmoc-$(V)_2$OMe, PhAc-$(F)_2$OMe, Cbz-$(F)_2$OMe, Nap-$(F)_2$OMe, Nap-(FY)OMe, Fmoc-(FY)OMe	Fmoc-$(L)_2$, Fmoc-$(L)_3$, Fmoc-$(V)_2$, PhAc-$(F)_2$, Cbz-$(F)_2$Nap-$(F)_2$, Nap-(FY), Fmoc-(FY)	Via bond cleavage	[22]
Thermolysin	Fmoc-A, Fmoc-V, Fmoc-L, Fmoc-F and $(F)_2$ or $(L)_2$, Fmoc-S and F-OMe, L-OMe, Fmoc-T and F-OMe, L-OMe	Fmoc-tripeptide	Via bond formation	[21, 23, 69]

4 Kinetics and Thermodynamics of Enzymatic Supramolecular Polymerisations

In terms of thermodynamics, two situations may arise in biocatalytic self-assembly: (1) the enzyme reaction and self-assembly process are both favoured in isolation, or (2) the enzyme reaction itself is thermodynamically unfavoured but involves a small change in free energy that can be overcome by the overall free energy change from the stabilisation of the self-assembled structure. The latter gives rise to a fully reversible system that operates under thermodynamic control (Fig. 4).

4.1 Enzyme-Triggered Assembly: Uncoupled

Uncoupled systems operate under kinetic control and allow access to kinetically trapped supramolecular structures (i.e. structures that do not represent the global free energy minimum), which can be achieved by simply altering the amount of catalyst present, allowing for enzymatic control of emergent properties. Such differential control of supramolecular polymerisation rate directly affects the viscoelastic properties of the resulting gel network. Rheology studies have shown that rapid formation of the self-assembled structure induces more mechanical strength to the hydrogel matrix than does a structure that is formed more slowly, with variations of orders of magnitude reported (Table 2).

The ability to control mechanical properties enzymatically has been demonstrated in two recent examples based on phosphatase-triggered self-assembly of aromatic peptide amphiphiles [55, 56]. A very simple system based on Fmoc-Y

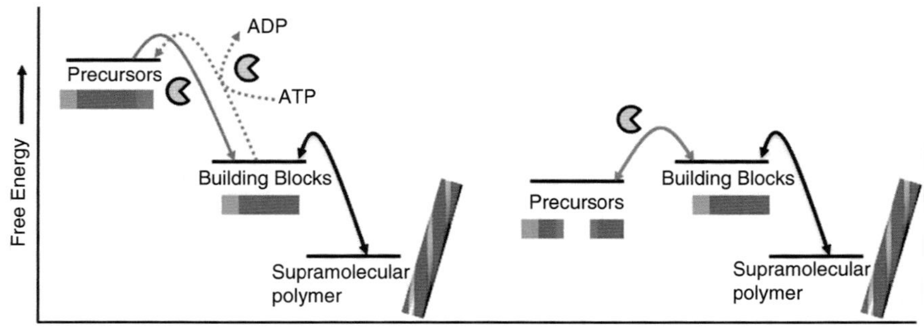

Fig. 4 Free energy diagram for the two possible situations in enzyme-triggered formation of supramolecular assembly. *Left*: The enzyme-catalysed reaction and self-assembly process are both favoured independently and therefore uncoupled. *Right*: Enzyme-triggered self-assembly under thermodynamic control; formation of the building blocks is thermodynamically unfavoured in isolation and occurs in reversible fashion when coupled to a sufficiently stable self-assembled structure formation

Table 2 Effects of alkaline phosphatase concentration on the gelation time and mechanical properties of the peptide hydrogels of Fmoc-Y-OH (entries 1–4) and the β-peptide hydrogel (entries 5–7)

Entry	Concentration of enzyme (U mL^{-1})	Time required for hydrogelation (min)	G′ (kPa)
1	10	60–120	34 ± 13
2	3	60–120	16 ± 7
3	1	120–240	9 ± 8
4	0.1	>1440	12 ± 6
5	5.88	2	4
6	2.94	10	0.9
7	1.47	30	0.3

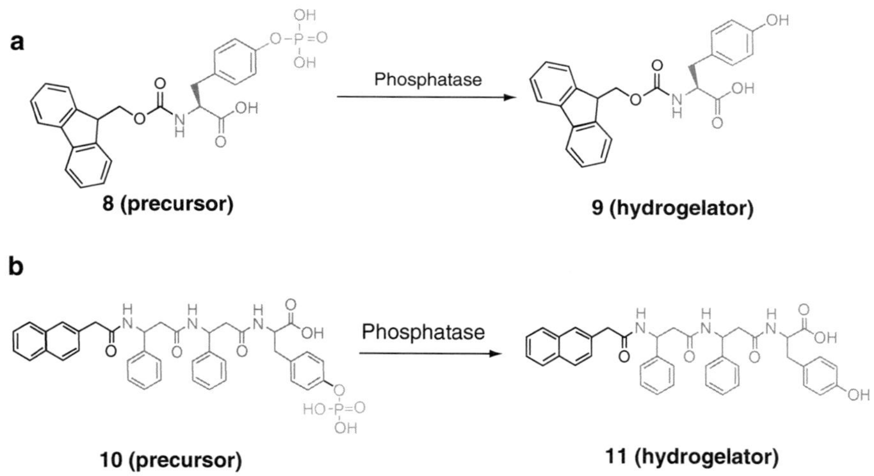

Fig. 5 (**a**) Dephosphorylation of Fmoc-Y(*p*)-OH (**8**) to hydrogelator Fmoc-Y-OH (**9**) by alkaline phosphatase. (**b**) Conversion of the β-peptide precursor **10** to hydrogelator **11** using acid phosphatase at pH 4.8

demonstrated a threefold increase in gel stiffness upon increase of enzyme concentration by two orders of magnitude (Fig. 5a, Table 2, entries 1–4).

A second example utilised a β-peptide derivative (Fig. 5b). The elastic modulus could be varied from 4000 to 300 Pa (Table 2, entries 5–7), thus indicating that the ratio of enzyme to precursor plays a significant role in tailoring the supramolecular assemblies.

Such control over the mechanical strength of the supramolecular gel network is of particular interest for applications in cell culture, particularly in stem cell research because the elastic modulus of cell substrates has been found to dictate the differentiation pathways chosen by mesenchymal stem cells [57].

4.2 Enzyme-Triggered Self-Assembly Under Thermodynamic Control

Proteases are well known for their ability to hydrolyse peptide bonds. The Gibbs free energy change of amide synthesis/hydrolysis is small and the reaction is readily reversed, for example by relative stabilisation of the peptide over the hydrolysis products, as shown previously in organic solvent systems and heterogeneous reaction media [58–60]. By coupling protease catalysis with peptide self-assembly, a similar reversal of hydrolysis to preferred synthesis occurs whereby the self-assembly provides the thermodynamic driving force for peptide synthesis [21, 23]. The resulting thermodynamically controlled systems are fully reversible and will proceed towards the lowest accessible free energy state. Interestingly, when mixtures of starting materials are supplied, these systems should self-select the most thermodynamically stable structures from dynamic mixtures, as discussed below.

Dynamic combinatorial libraries (DCLs) are continuously interconverting libraries that eventually evolve to an equilibrium distribution [61–65]. This approach has been used successfully in the discovery of stable supramolecular assemblies from mixtures. Due to the nearly endless possible peptide sequences that can potentially be synthesised, the DCL approach is attractive for the identification of supramolecular peptide interactions. Indeed, disulfide exchange between cysteine residues has been explored for this purpose [66, 67] as has peptide–metal binding [68]. We have recently demonstrated protease-catalysed amide exchange in this context, which allows for the evolution of the self-assembled peptide structures, and will therefore allow exploration of peptide sequence space for biomaterials design.

Evolution of peptide nanostructures has been investigated for Fmoc-L peptides (Fig. 6). When Fmoc-L and LL were exposed to a non-selective protease, a Fmoc-L_n oligomer distribution results. Upon initiation of the reaction, Fmoc-L_3 is found to be formed as the major component (because it is the direct coupling product between the starting materials). Overtime, the system rearranges itself and eventually reaches an equilibrium distribution in which Fmoc-L_5 is predominant (Fig. 6). Analysis by atomic force microscopy (AFM) showed a drastic change in morphology from fibres (Fmoc-L_3) to sheet-like structures (Fmoc-L_5) (Fig. 6a) [21] and suggests that the sheet-like pentapeptide structure represents the lowest accessible folded state for this system. This enzymatic DCL approach was also explored for the screening of a range of dipeptide sequences in Fmoc-dipeptide-methyl ester gelators [69].

Thus, dynamic peptide libraries offer the potential to identify the most stable self-assembled supramolecular nanostructure from a mixture of several components. This opens up the possibility of exploiting the versatility of peptides for the discovery of new nanostructures.

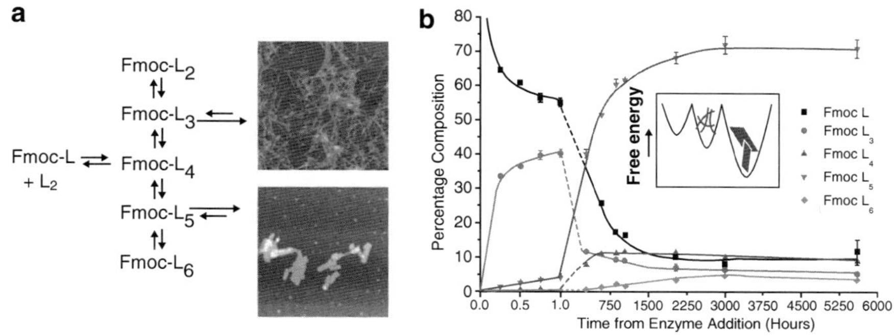

Fig. 6 Dynamic combinatorial peptide library that exploits enzyme reactions to control self-assembly processes under thermodynamic control. (**a**) Emergence of the potential peptide derivatives of varying length in a library of interconverting molecules formed from the staring materials of Fmoc L/L_2 system. Fmoc-L_5 is preferentially formed. Corresponding AFM images of the fibrillar structures formed at 5 min after the addition of enzyme, and the sheet-like structures observed after 2000 h show that redistribution of the derivatives is accompanied by the remodelling from fibres (Fmoc L_3) to sheet-like structures (Fmoc L_5). (**b**) HPLC analysis of the composition of the system reveals the formation and the stabilisation of Fmoc-L_5 over time. Modified from [21]

5 Spatiotemporal Control of Nucleation and Growth in Enzymatic Supramolecular Polymerisation

In supramolecular polymerisations, it is a major challenge to gain sufficient control over the nucleation and the structure growth as these are generally bulk processes that are controlled by overall changes in reaction conditions such as pH, ionic strength, temperature, etc. Enzymatic supramolecular polymerisations offer unique opportunities in this context. Depending on the relative rates of enzyme reaction and diffusion of the building blocks near an active enzyme, it is possible to achieve a local concentration above the critical aggregation that favours structure formation. For example, microscopic analysis of the initial stages of the self-assembly process has shown that the fibres propagate from spherical structures of approximately 30 nm in diameter (Fig. 7a, b). Because enzymatic formation of the building blocks is localised, it appears that self-assembly nucleates close to enzyme molecules. This suggests that each sphere contains a small number of enzyme molecules from which fibres propagate over time [21].

Spatially confined self-assembly has been further confirmed by localizing thermolysin on certain areas of a PEGylated surface. Upon immersion of this modified surface into a solution containing self-assembling precursors, nanostructures were formed in the vicinity of the enzyme, as observed through congo-red staining (Fig. 7c) [21]. Thus, enzyme-assisted self-assembly allows for construction of supramolecular polymers with spatiotemporal control, i.e. where and when they are required.

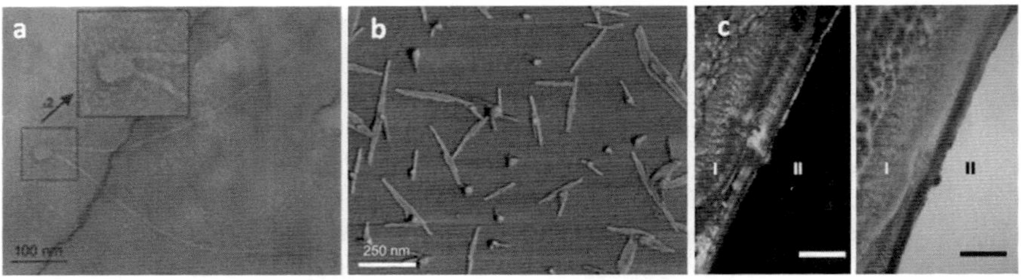

Fig. 7 Spatial confinement of nucleation and growth as observed for Fmoc-L/L_2 system. (**a**) TEM image of typical features observed with fibre propagation demonstrates confined fibre growth from spherical structures. (**b**) The same process is visualised by AFM. (**c**) *Right*: Localised self-assembly visualised by congo-red staining under cross-polarised light. Thermolysin is coupled to the glass surface through PEG diamine and glutaraldehyde crosslinking. Upon immersion into the solution of Fmoc-L/L_2 solution, self-assembly is observed in the vicinity of the immobilised enzyme. *Left*: Association of the congo-red dye with the β-sheet of the fibres results in green birefringence (*I*), no effect is observed on the regions without enzyme (*II*). Modified from [21]

6 Applications in Biomedicine and Nanotechnology

Supramolecular polymers are finding extensive applications in several areas of biomedicine, which include 3D cell culture, targeted drug delivery, biosensing, inhibitor screening and wound healing. For example, self-assembled hydrogelating systems of aromatic peptide amphiphiles comprising a fibrous network are suitable as minimal mimics of the extracellular matrix and have been used to support 2D and 3D culture of cells [10, 70, 71]. Using enzyme action to control formation of these gels would provide enhanced control over stiffness, topography and chemical composition and may give rise to next generation scaffolds for the controlled differentiation of stem cells.

An interesting recent example in the context of drug delivery of such enzyme-triggered self-assembly includes the modification of the anticancer drug taxol (Fig. 8) into a self-delivering supramolecular gel. The drug is attached covalently to a motif that forms a self-assembled structure, which bears a molecular switch component that is cleavable by a phosphatase (Fig. 8) [72]. On exposure to alkaline phosphatase, the phosphate group of **12** is hydrolysed to give rise to **13** that forms a hydrogel. This hydrogel can slowly release the taxol derivative **13** in aqueous solution when it is exposed to phosphate-buffered saline. This derivative showed similar activity to taxol in toxicology studies. Thus, this enzyme-triggered bioactive hydrogel generates a facile strategy of dissolving otherwise-insoluble hydrophobic drugs in aqueous solution. This example proves that small molecule drugs are excellent candidates for engineering functional hydrogels without compromising their activities.

Enzyme-sensitive supramolecular polymers also hold promise in analytical applications such as the screening of enzyme inhibitors. A simple visual assay based on the hydrogelation of small molecules has been developed for screening the activities of inhibitors of enzymes like acid phosphatase. A number of inhibitors for

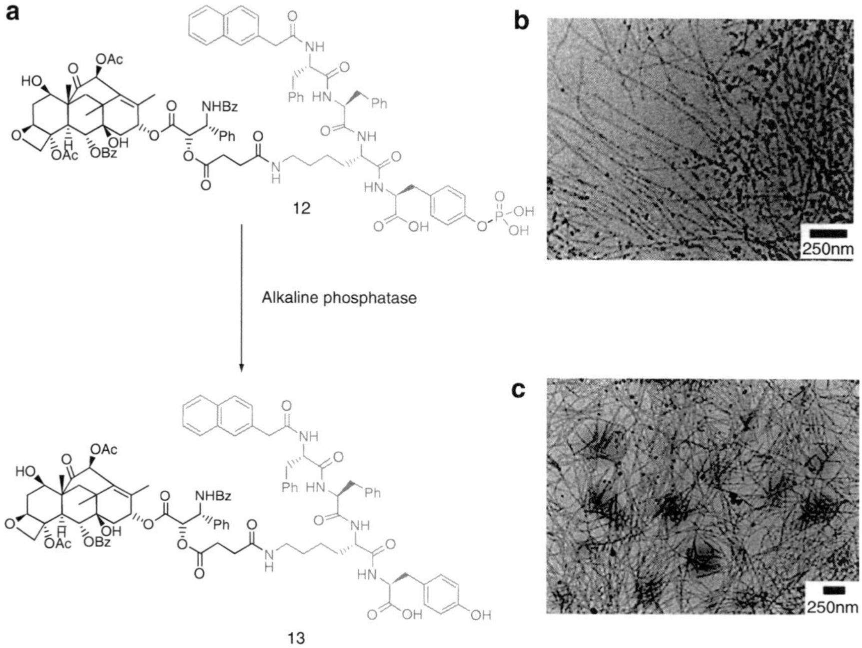

Fig. 8 (**a**) Modification of taxol with a self-assembly motif containing a molecular switch, phosphate (**12**), cleavable by phosphatase. On exposure to alkaline phosphatase, this forms **13**, which forms a hydrogel. (**b**) TEM image of the solution of **12** at 5 min after the addition of alkaline phosphatase. (**c**) TEM image of the hydrogel of **13**, overnight after the addition of alkaline phosphatase. Figure Adapted from [72] Copyright 2009 American Chemical Society

acid phosphatase, such as pamidronate disodium, Zn^{2+}, and sodium orthovanadate (Na_3VO_4), have been exploited to estimate their minimum inhibition concentrations [73].

The formation of dynamic artificial supramolecular nanoscale assemblies within living cells will be of interest to find out its influence on cell behaviour since it will integrate molecular self-assembly with biochemical reactions, giving rise to better understanding of biological processes and possible new approaches to tackle diseased states.

To exploit enzymatic hydrogelation intracellularly, Bing Xu and his team designed a non-assembling peptide precursor molecule (**14**). When exposed to the cells, it enters into the cell by diffusion and, being an esterase substrate, it undergoes hydrolysis by an endogenous esterase enzyme to form **15**, which self-assembles into nanostructures (Fig. 9) [74, 75]. This gelation induces an abrupt change in the viscosity of the cell cytoplasm and cause cell death, which depends on enzyme expression levels in the cells under investigation. At a certain concentration of **14** (0.08 wt%), the majority of cells derived from human cervical cancer tumours died within 3 days, whereas fibroblast NIH3T3 cells remained alive under the same conditions (Fig. 9b).

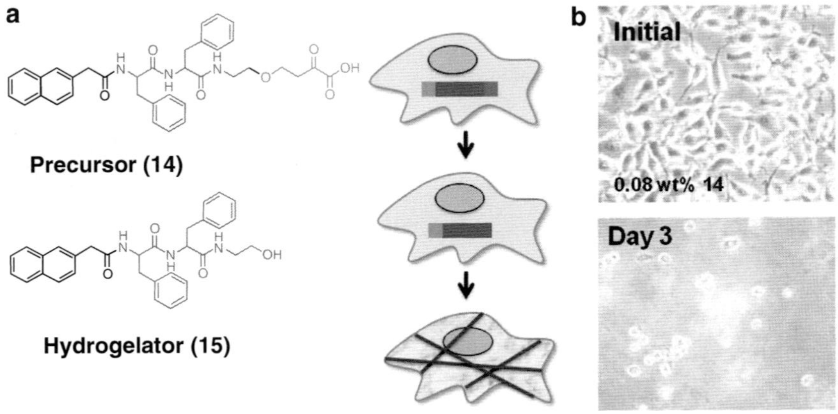

Fig. 9 Intracellular cleavage of an esterase substrate (**14**) by an endogenous esterase to the hydrogelator (**15**) and corresponding formation of supramolecular assembly within the cells. Reproduced with permission from [75] Copyright Wiley-VCH Verlag GmbH & Co. KGaA

7 Conclusion and Outlook

A number of dynamic supramolecular polymers control vital functions in biology. These are tightly regulated by highly selective and spatially confined catalytic mechanisms whereby non-assembling precursors are catalytically activated to produce self-assembling components.

Over the past 5 years, a number of researchers have started to explore and mimic these approaches in the laboratory. Enzyme-assisted formation of supramolecular polymers has several unique features. These include selectivity, confinement and catalytic amplification, which allow for superior control as observed in biological systems. These systems are finding applications in areas where supramolecular function is directly dictated by molecular order, for example in designed biomaterials for 3D cell culture, templating, drug delivery, biosensing, and intracellular polymerisations to control cell fate. Overall, biocatalytic production of supramolecular polymers provides a powerful new paradigm in stimuli-responsive nanomaterials.

References

1. Lehn JM (1995) Supramolecular chemistry – concepts and perspectives. VCH Weinheim
2. Lehn JM (2002) Supramolecular polymer chemistry – scope and perspectives. Polym Int 51:825–839
3. Mart RJ, Osborne RD, Stevens MM, Ulijn RV (2006) Peptide-based stimuli-responsive biomaterials. Soft Matter 2:822–835
4. Jayawarna V, Ali M, Jowitt TA, Miller AF, Saiani A, Gough JE, Ulijn RV (2006) Nanostructured hydrogels for three-dimensional cell culture through self-assembly of fluorenylmethoxycarbonyl-dipeptides. Adv Mater 18:611–614

5. Silva GA, Czeisler C, Niece KL, Beniash E, Harrington DA, Kessler JA, Stupp SL (2004) Selective differentation of neural progenitor cells by high-epitope density nanofibers. Science 303:1352–1355
6. Holmes TC, de Lacalle S, Su X, Liu G, Rich A, Zhang S (2000) Extensive neurite outgrowth and active synapse formation on self-assembling peptide scaffolds. Proc Natl Acad Sci 97:6728–6733
7. Haines LA, Rajagopal K, Ozbas B, Salick DA, Pochan DJ, Schneider JP (2005) Light-activated hydrogel formation via the triggered folding and self-assembly of a designed peptide. J Am Chem Soc 127:17025–17029
8. Matsumoto S, Yamaguchi S, Ueno S, Komatsu H, Ikeda M, Ishizuka K, Iko Y, Tabata KV, Aoki H, Ito S, Noji H, Hamachi I (2008) Photo gel-sol/sol-gel transition and its patterning of a supramolecular hydrogel as stimuli-responsive. Biomaterials 14:3977–3986
9. Muraoka T, Cui H, Stupp SI (2008) Quadruple helix formation of a photoresponsive peptide amphiphile and its light-triggered dissociation into single fibers. J Am Chem Soc 130:2946–2947
10. Yang Z, Liang G, Xu B (2008) Enzymatic hydrogelation of small molecules. Acc Chem Res 41:315–326
11. Ulijn RV (2006) Enzyme responsive materials: a new class of smart biomaterials. J Mat Chem 16:2217–2225
12. Colombo M, Brittingham RJ, Klement JF, Majsterek I, Birk DE, Uitto J, Fertala A (2003) Procollagen VII self-assembly depends on site-specific interactions and is promoted by cleavage of the NC2 domain with procollagen C-proteinase. Biochemistry 42:11434–11442
13. Kessler E, Takahara K, Biniaminov L, Brusel M, Greenspan DS (1996) Bone morphogenetic protein-1: the type I procollagen C-proteinase. Science 271:360–362
14. Prockop DJ, Kivirikko KI (1995) Collagens: molecular biology, diseases, and potentials for therapy. Annu Rev Biochem 64:403–434
15. Feratala A, Sieron A, Hojima Y, Ganguli A, Prockop DJ (1994) Self-assembly into fibrils of collagen II by enzymic cleavage of recombinant procollagen II. Lag period, critical concentration, and morphology of fibrils differ from collagen I. J Biol Chem 269:11584–11589
16. Kadler K, Hojima Y, Prockop DJ (1987) Assembly of collagen fibrils de novo by cleavage of the type I pC-collagen with procollagen C-proteinase. Assay of critical concentration demonstrates that collagen self-assembly is a classical example of an entropy-driven process. J Biol Chem 262:15696–15701
17. Lodish H, Berk A, Matsudaira P, Kaiser CA, Krieger M, Scott, MP, Zipursky SL, Darnell J (2003) Molecular cell biology, 5th edn. Freeman, New York
18. Wu L.-Q, Payne GF (2004) Biofabrication: the use of biological materials and biocatalysts to construct nanostructured assemblies. Trends Biotechnol 22:593–599
19. Hu BH, Messersmith PB (2003) Rational design of transglutaminase substrate peptides for rapid enzymatic formation of hydrogels. J Am Chem Soc 125:14298–14299
20. Sperinde JJ, Griffith LG (1997) Synthesis and characterization of enzymatically-cross- linked poly(ethylene glycol) hydrogels. Macromolecules 30:5255–5264
21. Williams RJ, Smith AM, Collins R, Hodson N, Das AK, Ulijn RV (2009) Enzyme- assisted self-assembly under thermodynamic control. Nat Nanotechnol 4:19–24
22. Das AK, Collins R, Ulijn RV (2008) Exploiting enzymatic (reversed) hydrolysis in directed self-assembly of peptide nanostructures. Small 4:279–287
23. Toledano S, Williams RJ, Jayawarna V, Ulijn RV (2006) Enzyme-triggered self-assembly of peptide hydrogels via reversed hydrolysis. J Am Chem Soc 128:1070–1071
24. Yang Z, Ma M, Xu B (2009) Using matrix metalloprotease-9 (MMP-9) to trigger supramolecular hydrogelation. Soft Matter 5:2546–2548
25. Adler-Abramovich L, Perry R, Sagi A, Gazit E, Shabat D (2007) Controlled assembly of peptide nanotubes triggered by enzymatic activation of self-immolative dendrimers. Chembiochem 8:859–862
26. Dos Santos S, Chandravarkar A, Mandal B, Mimna R, Murat K, Saucede L, Tella P, Tuchscherer G, Mutter M (2005) Switch-peptides: controlling self-Assembly of amyloid β-derived peptides in vitro by consecutive triggering of acyl migrations. J Am Chem Soc 127:11888–11889

27. Yang Z, Gu H, Fu D, Gao P, Lam JK, Xu B (2004) Enzymatic formation of supramolecular hydrogels. Adv Mater 16:1440–1444
28. Winkler S, Wilson D, Kaplan DL (2000) Controlling beta-sheet assembly in genetically engineered silk by enzymatic phosphorylation/dephosphorylation. Biochemistry 39:12739–12746
29. Kühnle H, Börner HG, (2009) Biotransformation on polymer-peptide conjugates: a versatile tool to trigger microstructure formation. Angew Chem Int Ed 48:6431–6434
30. Amir RJ, Zhong S, Pochan DJ, Hawker CJ (2009) Enzymatically triggered self-assembly of block copolymers. J Am Chem Soc 131:13949–13951
31. Reches M, Gazit E (2003) Casting metal nanowires within discrete self-assembled peptide nanotubes. Science 300:625–627
32. Plunkett KN, Berkowski KL, Moore JS (2005) Chymotrypsin responsive hydrogel: application of a disulfide exchange protocol for the preparation of methacrylamide containing peptides Biomacromolecules 6:632–637
33. Lutolf MP, Raeber GP, Zisch AH, Tirelli N, Hubbell JA (2003) Cell-responsive synthetic hydrogels. Adv Mater 15:888–892
34. Van Bommel KJC, Stuart MCA, Feringa BL, Van Esch J (2005) Two-stage enzyme mediated drug release from LMWG hydrogels. Org Biomol Chem 3:2917–2920
35. Jun HW, Yuwono V, Paramonov, SE, Hartgerink JD (2005) Enzyme-mediated degradation of peptide-amphiphile nanofiber networks. Adv Mater 17:2612–2617
36. Chau Y, Luo Y, Cheung ACY, Nagai Y, Zhang SG, Kobler JB, Zeitels SM, Langer R (2008) Incorporation of a matrix metalloproteinase-sensitive substrate into self-assembling peptides as a model for biofunctional scaffolds. Biomaterials 29:1713–1719
37. Yang Z, Liang G, Wang L, Xu B (2006) Using a kinase/phosphatase switch to regulate a supramolecular hydrogel and forming the supramoleclar hydrogel in vivo. J Am Chem Soc 128:3038–3043
38. Um SH, Lee JB, Park N, Kwon SY, Umbach CC, Luo D (2006) Enzyme-catalysed assembly of DNA hydrogel. Nat Mater 5:797–801
39. Zhang SG (2003) Fabrication of novel biomaterials through molecular self-assembly. Nat Biotechnol 21:1171–1178
40. Whitesides GM, Grzybowski BA (2002) Self-assembly at all scales. Science 295:2418–2421
41. Grzybowski BA, Wilmer CE, Kim J, Browne KP, Bishop KJM (2009) Self-assembly: from crystals to cells. Soft Matter 5:1110–1128
42. Hirst AR, Coates IA, Boucheteau TR, Miravet JF, Escuder B, Castelletto V, Hamley IW, Smith DK (2008) Low-molecular-weight gelators: elucidating the principles of gelation based on gelator solubility and a cooperative self-assembly model. J Am Chem Soc 130:9113–9121
43. Rughani RV, Schneider JR (2008) Molecular design of beta-hairpin peptides for material construction. MRS Bull 33:530–535
44. de Loos M, Feringa BL, van Esch JH (2005) Design and application of self-assembled low molecular weight hydrogels. Eur J Org Chem 3615–3631
45. Kobayashi H, Friggeri A, Koumoto K, Amaike M, Shinkai S, Reinhoudt DN (2002) Molecular design of "super" hydrogelators: understanding the gelation process of azobenzene-based sugar derivatives in water. Org Lett 4:1423–1426
46. Woolfson DN, Ryadnov MG (2006) Peptide-based fibrous biomaterials: some things old, new and borrowed. Curr Opin Chem Biol 10:559–567
47. Yanlian Y, Ulung K, Xiumei W, Horii A, Yokoi H, Zhang S (2009) Designer self-assembling peptide nanomaterials. Nano Today 4:193–210
48. Estroff LA, Hamilton AD (2004) Water gelation by small organic molecules. Chem Rev 104:1201–1218
49. Reches M, Gazit E (2006) Controlled patterning of aligned self-assembled peptide nanotubes. Nat Nanotechnol 1:195–200
50. Smith AM, Williams RJ, Tang C, Coppo P, Collins RF, Turner ML, Saiani A, Ulijn RV (2008) Fmoc-diphenylalanine self assembles to a hydrogel via a novel architecture based on π-interlocked β-sheets. Adv Mater 20:37–41
51. Smith AM, Ulijn RV (2008) Designing peptide based nanomaterials. Chem Soc Rev 37:664–675

52. Adams DJ, Butler MF, Frith WJ, Kirkland M, Mullen L, Sanderson P (2009) A new method for maintaining homogeneity during liquid–hydrogel transitions using low molecular weight hydrogelators. Soft Matter 5:1856–1862
53. Vegners R, Shestakova I, Kalvinsh I, Ezzell RM, Janmey PA (1995) Use of a gel-forming dipeptide derivative as a carrier for antigen presentation. J Pept Sci 1:371–378
54. Tang C, Smith AM, Collins RF, Ulijn RV, Saiani A (2009) Fmoc-diphenylalanine self-assembly mechanism induces apparent pK(a) shifts. Langmuir 25:9447–9453
55. Thornton K, Smith AM, Merry CLR, Ulijn RV (2009) Controlling stiffness in nanostructured hydrogels produced by enzymatic dephosphorylation. Biochem Soc Trans 37:660–664
56. Yang ZM, Liang GL, Xu B (2007) Enzymatic control of the self-assembly of small molecules: a new way to generate supramolecular hydrogels. Soft Matter 3:515–520
57. Engler AJ, Sen S, Sweeney HL, Discher DE (2006) Matrix elasticity directs stem cell lineage specification. Cell 126:677–689
58. Ulijn RV, De Martin L, Gardossi L, Halling PJ (2003) Biocatalysis with mainly undissolved solid substrates. Curr Org Chem 7:1333–1346
59. Klibanov AM (2001) Improving enzymes by using them in organic solvents. Nature 409:241–246
60. Halling PJ, Ulijn RV, Flitsch SL (2005) Understanding enzyme action on immobilized substrates. Curr Opin Biotechnol 16:385–392
61. Corbett PT, Leclaire J, Vial L, West KR, Wietor J-L, Sanders JKM, Otto S (2006) Dynamic combinatorial chemistry. Chem Rev 106:3652–3711
62. Rowan SJ, Cantrill SJ, Cousins GRL, Sanders JKM, Stoddart JF (2002) Dynamic covalent chemistry. Angew Chem Int Ed 41:898–952
63. Sreenivasachary N, Lehn JM (2005) Gelation-driven component selection in the generation of constitutional dynamic hydrogels based on guanine-quartet formation. Proc Nat Acad Sci 102:5938–5943
64. Otto S, Furlan RLE, Sanders JKM (2002) Selection and amplification of hosts from dynamic combinatorial libraries of macrocyclic disulfides. Science 297:590–593
65. Ludlow RF, Otto S (2008) Systems chemistry. Chem Soc Rev 37:101–108
66. Bilgiçer B, Xing X, Kumar K (2001) Programmed self-sorting of coiled coils with leucine and hexafluoroleucine cores. J Am Chem Soc 123:11815–11816
67. Krishnan-Ghosh Y, Balasubramanian S (2003) Dynamic covalent chemistry on self-templating peptides: formation of a disulfide-linked -hairpin mimic. Angew Chem Int Ed 42:2171–2173
68. Case MA, McLendon GL (2000) A virtual library approach to investigate protein folding and internal packing. J Am Chem Soc 122:8089–8090
69. Das AK, Hirst A, Ulijn RV (2009) Evolving nanomaterials using enzyme-driven dynamic peptide libraries (eDPL). Faraday Discuss 143:293–303
70. Zhou M, Smith AM, Das AK, Hodson NW, Collins RF, Ulijn RV, Gough JE (2009) Self-assembled peptide-based hydrogels as scaffolds for anchorage dependent cells. Biomaterials 30:2523–2530
71. Jayawarna V, Richardson SM, Hirst A, Hodson NW, Saiani A, Gough JE, Ulijn RV (2009) Introducing chemical functionality in Fmoc-peptide gels for cell culture. Acta Biomater 5:934–943
72. Gao Y, Kuang Y, Guo ZF, Guo Z, Krauss IJ, Xu B (2009) Enzyme-instructed self-assembly confers nanofibers and a supramolecular hydrogel of taxol derivative. J Am Chem Soc 131:13576–13577
73. Yang ZM, Xu B (2004) A simple visual assay based on small molecule hydrogels for detecting inhibitors of enzymes. Chem Commun 2424–2425
74. Yang ZM, Liang GL, Ma ML, Gao Y, Xu B (2007) In vitro and in vivo enzymatic formation of supramolecular hydrogels based on self-assembled nanofibers of a beta-amino acid derivative. Small 3:558–562
75. Yang ZM, Xu KM, Guo ZF, Guo ZH, Xu B (2007) Intracellular enzymatic formation of nanofibers results in hydrogelation and regulated cell death. Adv Mater 19:3152–3156

Index